IMAGES
of America

EARLY CRAIG

On the Cover: Craig Girls Tennis Club. Craig Girls Tennis Club members and friends are picnicking in June 1899. (Courtesy of the Museum of Northwest Colorado.)

IMAGES
of America

EARLY CRAIG

Museum of Northwest Colorado

ARCADIA
PUBLISHING

ISBN 978-0-7385-9969-4

Published by Arcadia Publishing
Charleston, South Carolina

Printed in the United States of America

Library of Congress Control Number: 2013935198

For all general information, please contact Arcadia Publishing:
Telephone 843-853-2070
Fax 843-853-0044
E-mail sales@arcadiapublishing.com
For customer service and orders:
Toll-Free 1-888-313-2665

Visit us on the Internet at www.arcadiapublishing.com

This book is dedicated to Glenda Cooper, assistant director of the Museum of Northwest Colorado from 1995 to 1999

Contents

ACKNOWLEDGMENTS

This book was written by Museum of Northwest Colorado director Dan Davidson, assistant director Janet Gerber, and registrar Mary Pat Dunn with assistance from staff members Rachel Van Tassel, Linda Herschberg, and Kristi Hankins. We would also like to thank those whose gracious donations have made our research library such an outstanding archive of regional pictorial history. The majority of the images in this book are from the Museum of Northwest Colorado archives unless otherwise noted. Sincere thanks goes to Janet Sheridan, who patiently edited the captions and the text for this book. The following entities generously lent us images for this book: Denver Public Library, Hayden Heritage Center, Tread of Pioneers Museum, and Wyman Living History Museum.

INTRODUCTION

The western slope of the Colorado Rocky Mountains perches at the edge of the Great Basin, which is the largest self-contained watershed in North America. The immense span between the Continental Divide and the Cascade and Sierra Mountain Ranges to the west falls away from the mountain peaks to semi-verdant valleys, arid basins, desert plains, and intermittent smaller mountain ranges. Not lush and arable when compared to Eastern United States, the region still offered fresh possibilities to the policymakers in Washington, DC, in the middle of the 19th century.

The Homestead Act of 1862 was enacted to push settlements beyond the Mississippi Valley. It was intended to open land in the West to small, independent farmers who had outgrown the populated eastern regions. The western states, though wildly varying in land features, hold similar historical progressions in common: the regional waves of homesteading hit about the same time and the end results of the homesteading era were similar. Economics, weather, and transportation are a few of the common forces that altered the outcome of homesteading throughout the Great Basin. Though there are stories of success and hardship unique to each area, those stories carry a unified theme that binds them together in the massive westward migration. Individual and family group characteristics along with challenges dictated by climate and land forms flavored each settlement. The homesteading period of the Yampa Valley reverberates with themes common to the entire West. But upon closer scrutiny, each little area also exudes its own individual temperament, twists of historical fate, and individual characteristics that color that town or locale with the distinctiveness and idiosyncrasies it possesses today.

From the beginning, agriculture, mining, and hunting were core activities of the little town first known as Yampa then later, Craig. Those three enterprises remain consistent themes in the life of the town and in its livelihood today. Founded in 1889, the townsite grew slowly but with characteristic western enthusiasm and parochial attitude that enabled the settlers to believe in their destiny as a viable community. Optimism was fueled by the belief that the little hamlet was on the envisioned route destined to link Denver with Salt Lake and thence the Pacific Northwest. Soon boasting a newspaper, saloon, hotel, school, and a church, the townspeople felt they were on their way to becoming a thriving center of commerce in the valley.

Geography and climate were not allies in the homesteaders' attempts to create a home in this remote region. The town was isolated in ways that many other settlements were not. It was cut off from Denver—the bustling Queen of the Plains—by the imposing peaks of the Rockies. The weather provided an additional challenge as it left the rough wagon roads leading into the area impassable much of the year. Getting supplies in and agricultural goods out was a challenge that often left the local populace fending for itself but at the same time instilled in the local people a can-do attitude and an innovative outlook that allowed them to survive the worst that weather could throw at them. The community is still, over a century later, marked by a self-sufficient attitude that has helped it survive numerous challenges.

In 1908, Craig was incorporated, and plans went forward to create infrastructures such as an electric power plant, sewer and water systems, and a marvelous amenity in the form of sidewalks. Still, the townspeople waited for the promised rail line, a dream that almost died in 1911 with the unexpected death of David Moffat, who had been the driving force behind the Moffat Road, as it was known. With a last push, the tracks into Craig were finally laid in the fall of 1913, and the first passenger train pulled in town in November of that year. The rails never extended any farther West, and the town had to be content with its status as a terminal point on the line. At least now, it had reliable transportation to the outside world.

The Enlarged Homestead Act of 1909 renewed interest in the homestead movement. With larger tracts of land available, more people began coming into the area, lured by the promise of free land and an escape from the industrialized cities of the East. These new homesteaders were often a different breed from those of the late 1880s. While the earlier wave had sent in families equipped with a heritage of farming and agriculture, these new homesteaders were often townspeople trying to reconnect with a lost rural heritage. By this time, most of the arable land had been settled, and this new generation of homesteaders, in some respects, faced a tougher challenge in their endeavors than their predecessors.

In 1911, due to its tremendous size, Routt County was divided into two counties. The western two-thirds was designated Moffat County, named after the great dreamer who had gambled his entire fortune to bring the railroad over the formidable Continental Divide to the little frontier town.

Craig continued to stretch itself to meet the demands of a growing population. By 1920, the town was developing its own sense of identity. With a new courthouse standing proudly on the main street, the community turned to building an armory, adding a new elementary school, and expanding its business district. An oil discovery in 1924 brought in new revenues and an industry that would become a vital part of the local economy for decades to come. The early years behind it, the town had developed into a plucky community with a firm vision for its future and a determination to continue to grow to its full potential.

One

Before Craig

The forested mountains, lush valleys, and arid plateaus of Northwestern Colorado are the starting point of the gradual descent of this Western watershed down to the waters of the Pacific Ocean. Before the arrival of the first Europeans to this area, a small nation of Utes roamed the land, harvesting wildlife and living an itinerant existence on the bounty provided in the wild beauties and extreme weathers found throughout the region.

The earliest documented exploration of the area was in 1776 by the Dominguez-Escalante Expedition. The men in this group were undoubtedly the first newcomers to view this section of the Colorado Plateau, including portions of today's Moffat County. Several decades later, spurred by the profitable beaver pelt trade, a few mountain men started trickling into the area. The region was also visited by important historical figures such as Davy Crocket, Jim Baker, Kit Carson, and John Fremont.

As the demand for western cattle gained strength in the mid-19th century, cattle operations filtered into the valleys of the Yampa River Basin, lured by the rich summer pastures of the open range. The arrival of the Union Pacific Railroad in Wyoming in 1869 paved the way for the next influx of newcomers. By the 1870s, the earliest settlers started finding their way into the region, and with them came the first permanent trappings of civilization.

As the first homesteaders arrived, tensions inevitably arose as they encroached upon the land historically inhabited by the Ute Nation. Flaring tempers and cultural misunderstandings culminated in the Meeker Massacre in which Indian agent Nathan Meeker and 10 of his employees were murdered by irate Utes. The outcome was the forcible removal of the Utes to a reservation in Utah.

That first era of settling in, before the town of Craig was established, is marked by a determined group of people who believed that a community in an area marked by extreme beauty as well as extreme weather could survive and flourish. Perhaps it is from these first pioneers that today's populace has inherited its dogged resolve to continue to build a life in a complex region of harsh challenges and rich natural splendors.

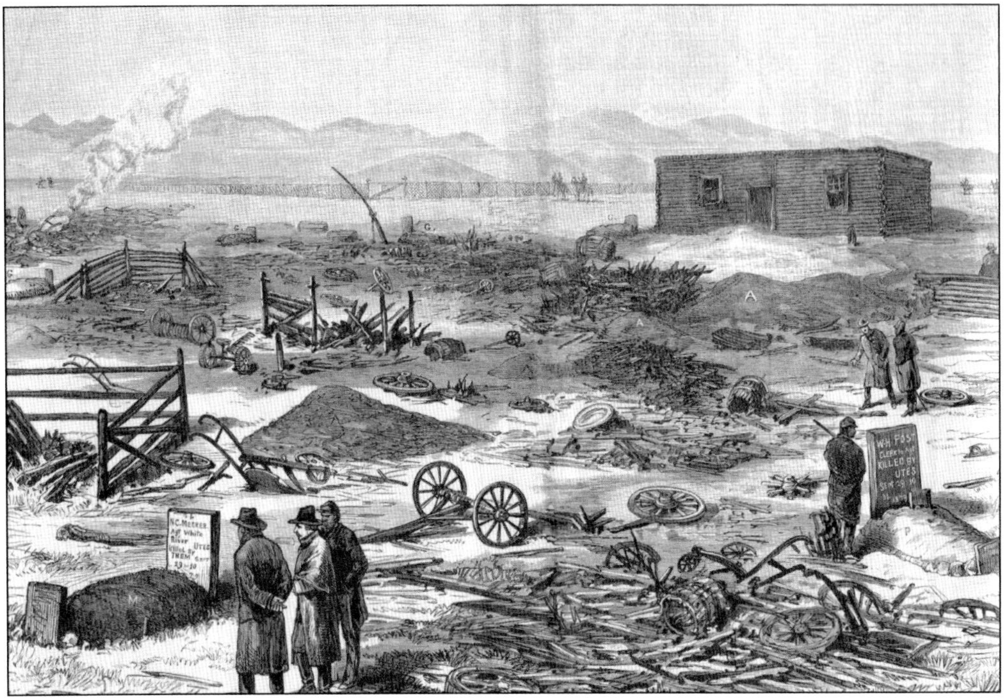

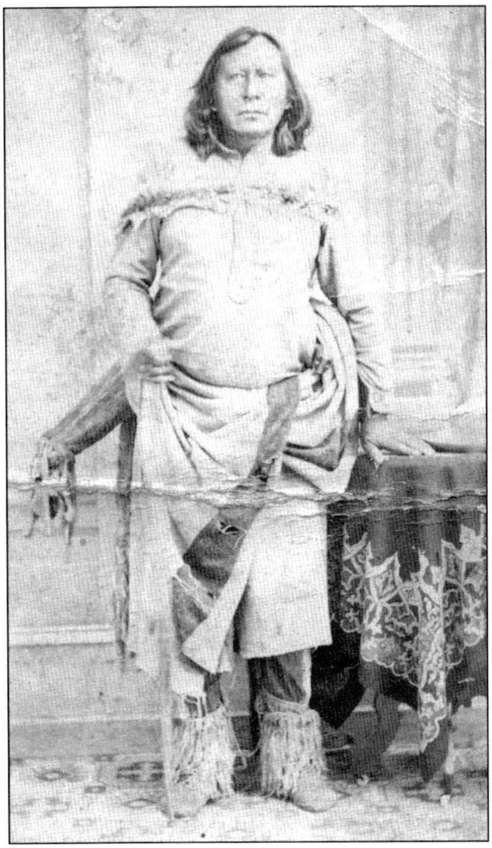

Aftermath of the Meeker Massacre, 1879. In September 1879, after Indian agent Nathan Meeker ordered the Utes' racetrack plowed up for farmland, the Utes revolted and killed Meeker and 10 of his employees. *Frank Leslie's Illustrated Newspaper* published this illustration of the scene after the uprising.

Chief Colorow. Born around 1810, Colorow developed a reputation as a warrior with strong leadership abilities. He led the northern band of Utes in clashes at Meeker and later, in the Battle of Milk Creek in 1879. Colorow again led in the Ute War of 1887 where it is thought he received a wound from which he never recovered. He died the following year.

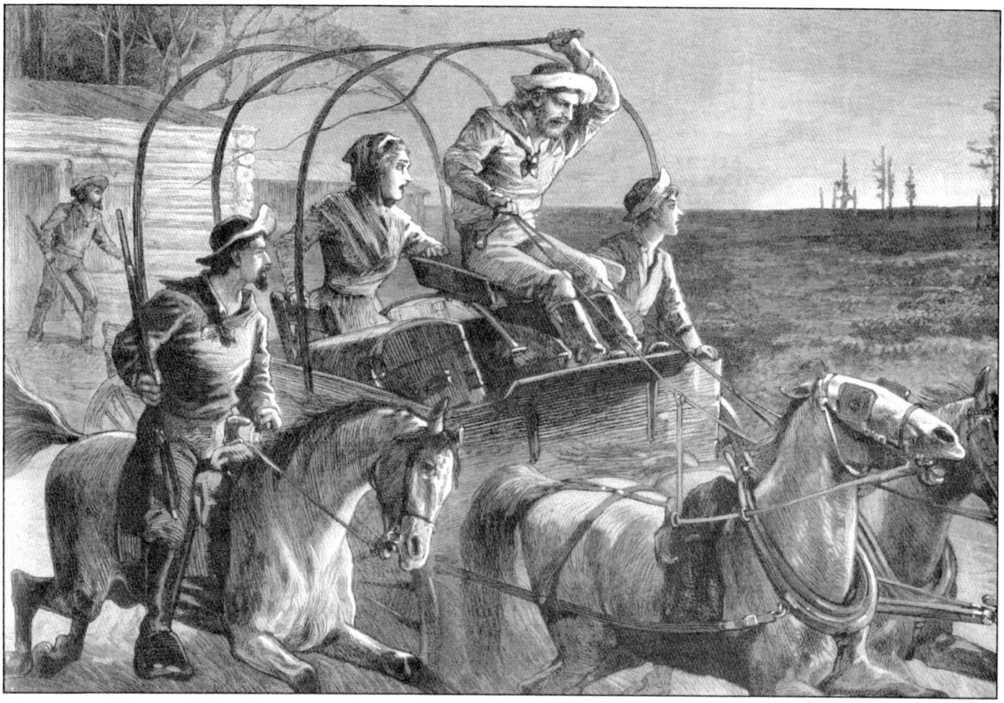

Fleeing a Feared Ute Attack, 1887. After Congressional hearings on the Meeker Massacre, the Ute Removal Act was passed, denying the Ute Nation 12 million acres of land it previously owned. The Utes were forced to move from the western slopes of the Rocky Mountains to Eastern Utah in 1881. With ongoing tensions, a skirmish over two stolen horses occurred in 1887. The resulting battle was reported as the "Ute War," which caused undue trepidation on the part of area settlers. The entire incident resulted in the death of two US soldiers, 10 Ute warriors, and a price tag of $80,000 for the US government. This image accompanied a story about that conflict in *Frank Leslie's Illustrated Newspaper.*

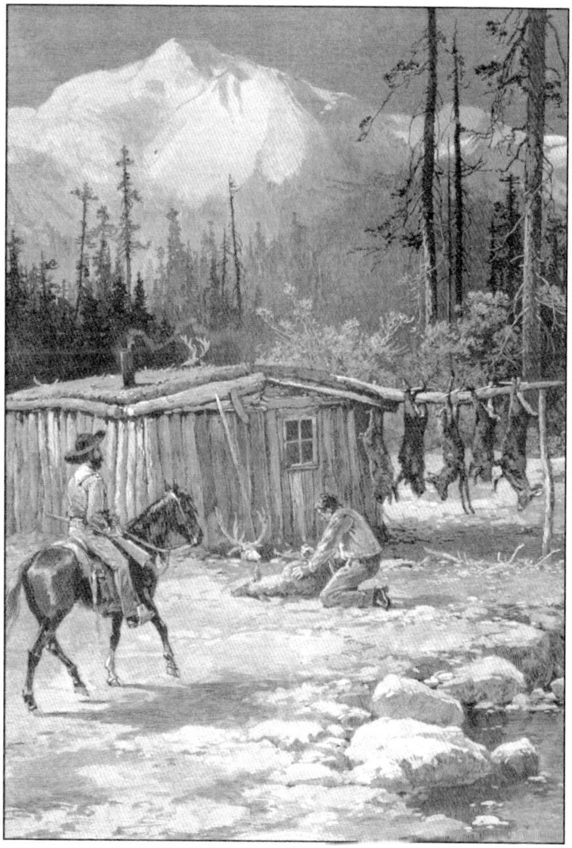

Trappers and Hunters in the American West. Fur trappers and, later, hunters were among the first people who came West during the 1830s to the 1860s. This illustration of a trapper's cabin is from *Frank Leslie's Illustrated Newspaper*, a literary and news magazine that often featured stories about the American West.

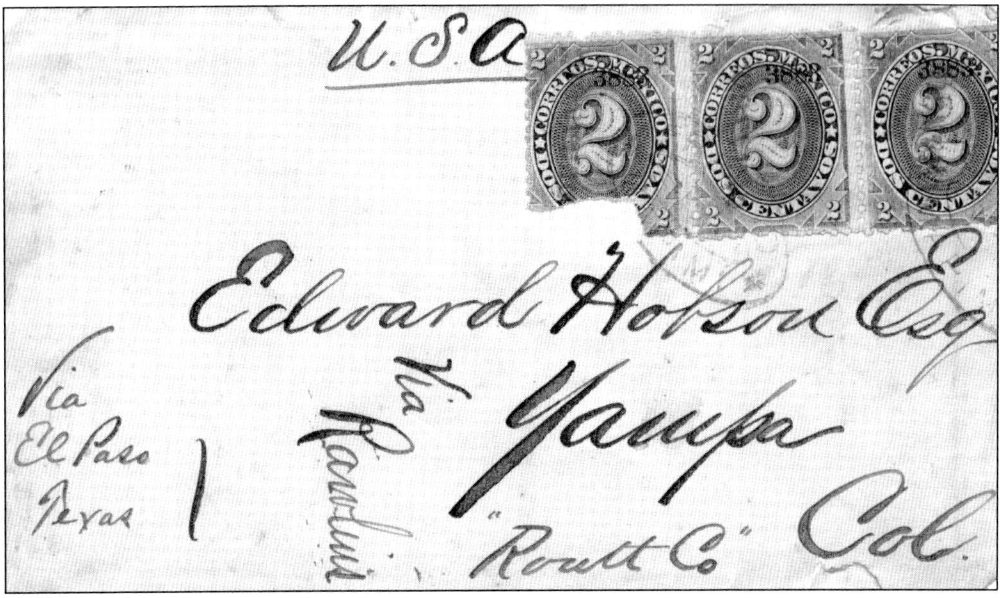

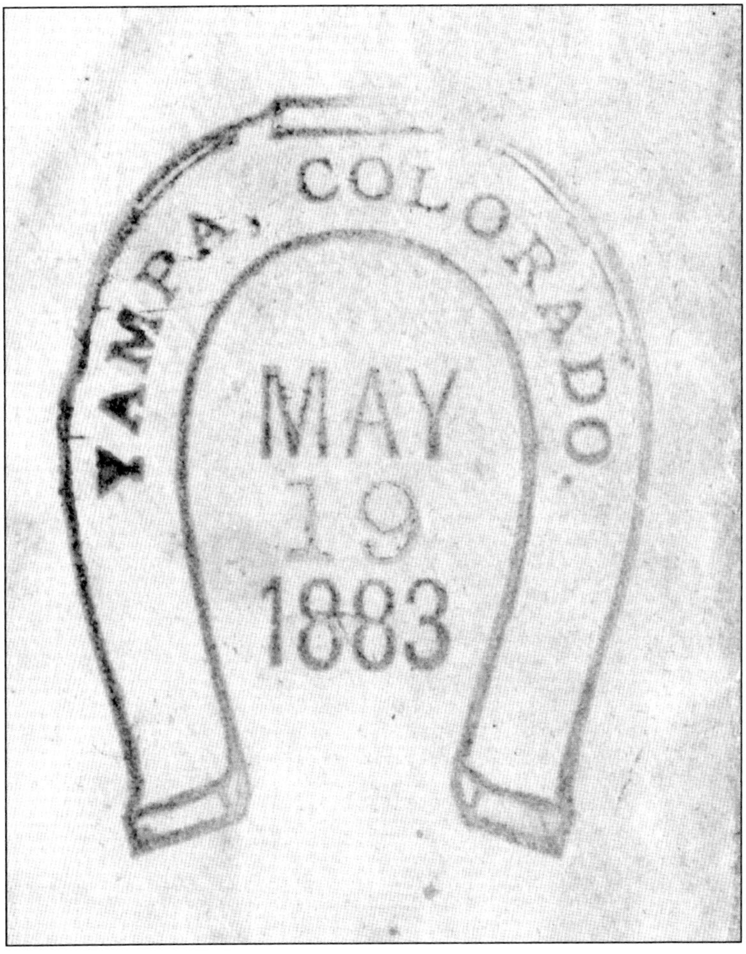

YAMPA, COLORADO POSTMARK, 1883. Yampa (not to be confused with present-day Yampa in Routt County) was Craig's second official post office name, the first name being Windsor. Windsor post office opened in 1877; its location moved several times over the next few years. "Yampa" was the second name for the area post office in 1880, and was located near the present-day county fairgrounds. This Yampa-postmarked envelope is dated 1883 and was originally mailed from Mexico, coming to Craig via Texas and then Rawlins.

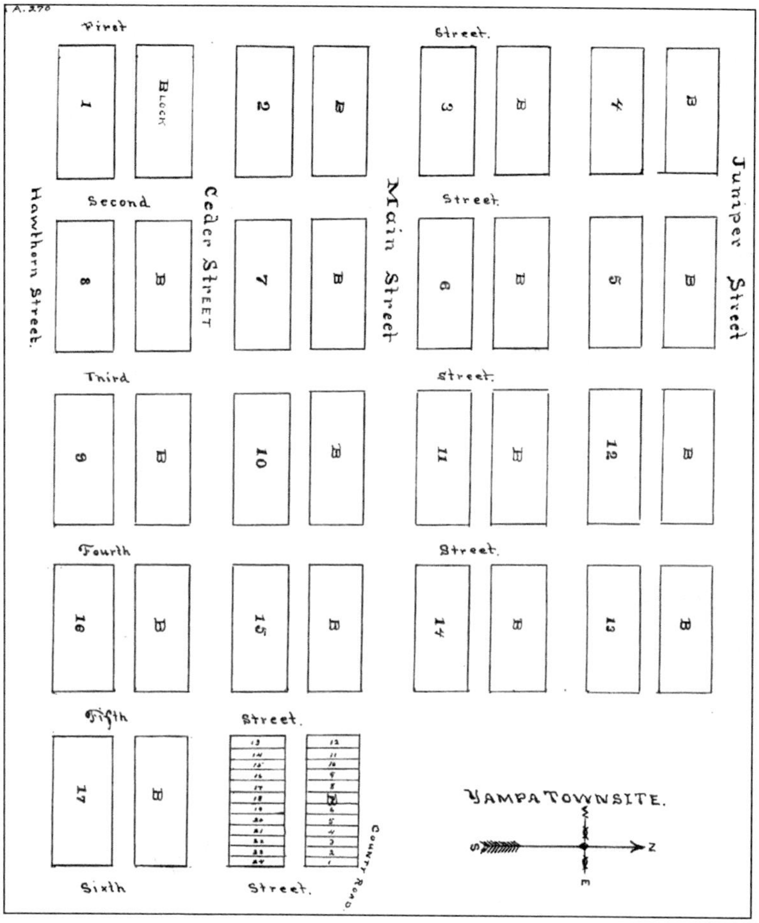

YAMPA PLAT. The original plat for the townsite of Yampa was filed in 1885.

HAHN'S PEAK, C. 1888. One of the earliest photographs of Hahn's Peak shows the first buildings in this mining town. The courthouse is on the right in the photograph. Here, a hunting party is seen with the village in the background. Before the Meeker Massacre in 1879, Hayden was the county seat, but just before the conflict, the county records were moved to Hahn's Peak. (Photograph by A.G. Wallihan.)

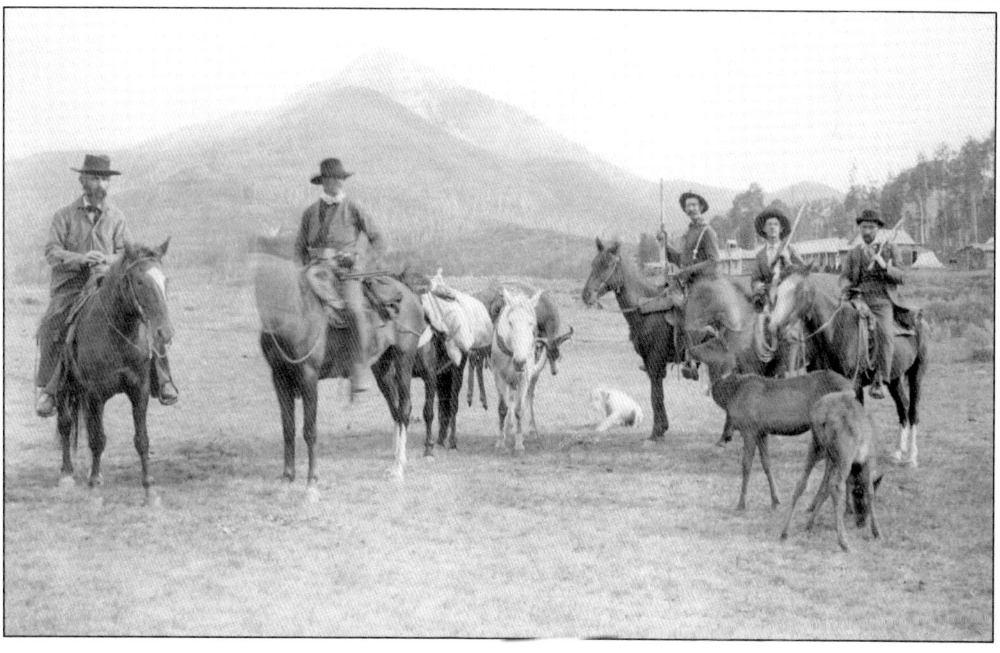

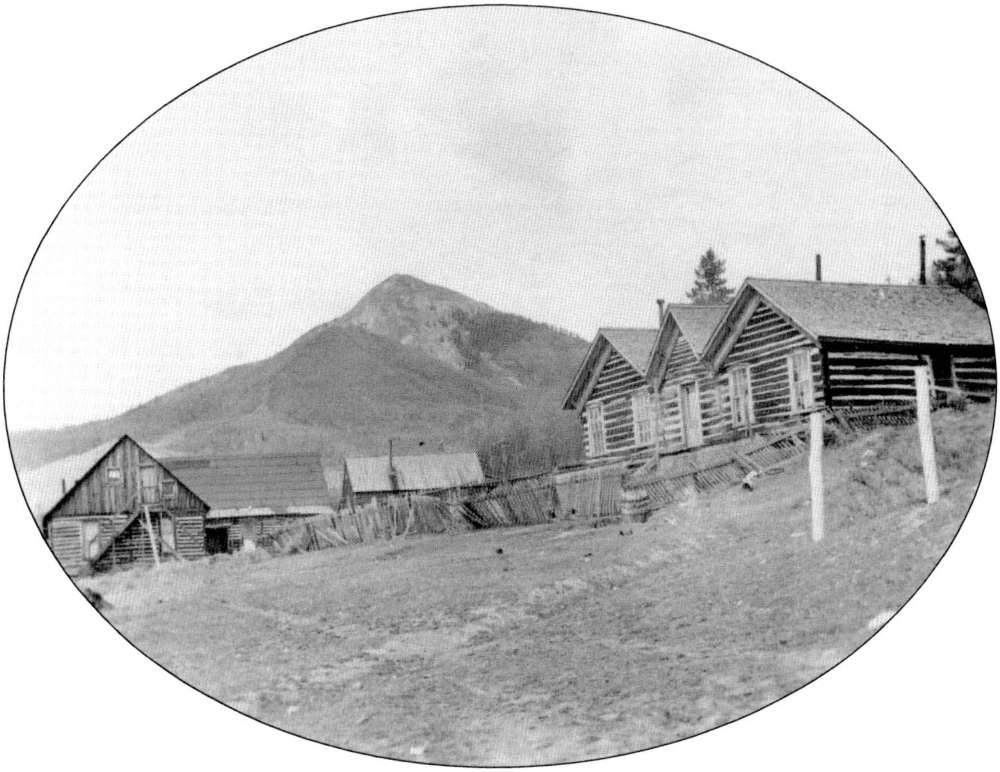

HAHN'S PEAK. For early residents of the Craig area (which, until 1911, was part of Routt County), transacting any official county or land business required an arduous trip of more than 90 miles one way to the remote county seat.

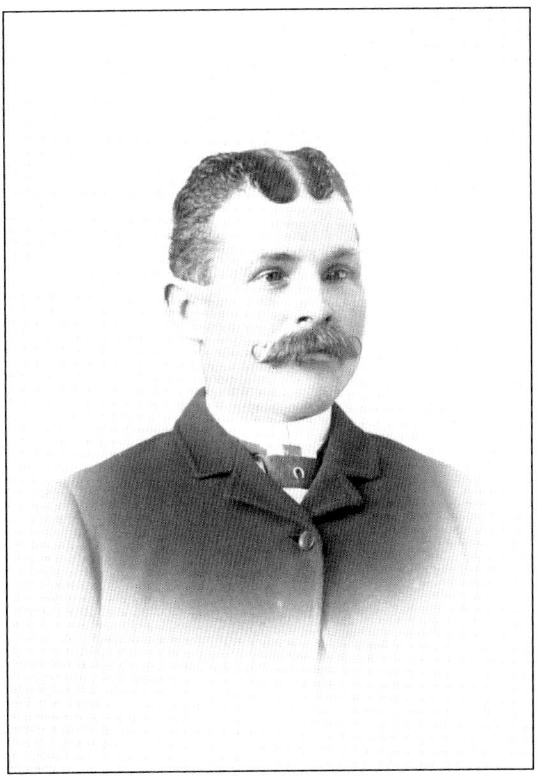

JOHN MACK, C. 1885. Born in Germany in 1857, John Mack immigrated to the United States and was one of the earliest settlers into Northwestern Colorado. He settled on land near the Yampa River southwest of town, where his descendants still live today.

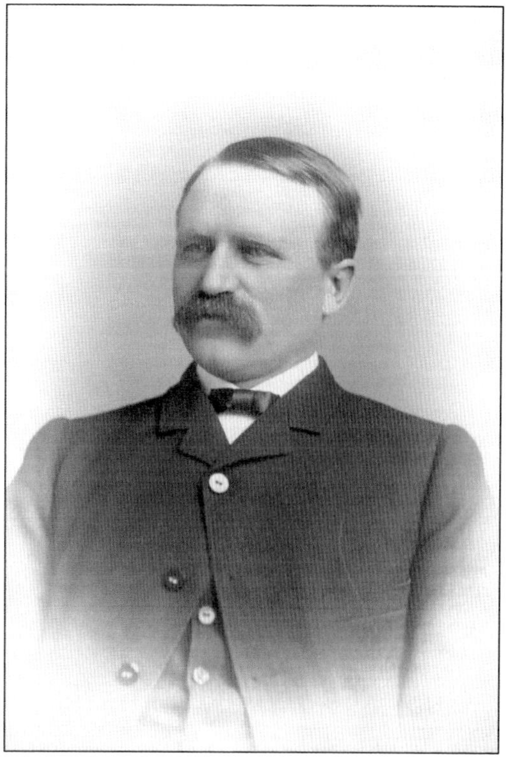

William H. Rose, 1892, and Rose Cabin, 1910. William H. Rose, born in 1844, was the first settler into what would become the Craig townsite around 1882. He built his cabin on the south side of present-day East Victory Way near Fortification Creek. Some of his homestead land was later purchased in 1889 by the Craig Land and Mercantile Company when that entity was laying out the proposed Craig townsite. Shortly after the townsite was founded, Rose started selling off additional parcels of his land in a subdivision known as Rosedale, which includes the residential area around present-day City Park. Some of Craig's earliest homes were built in this area. (Right, photograph by W.O. Luke; below, photograph by Robert M. Richardson IV.)

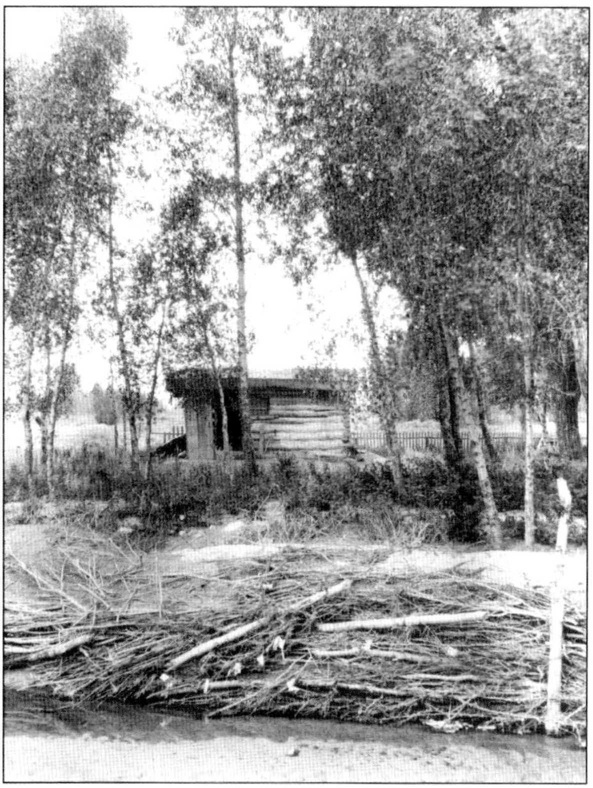

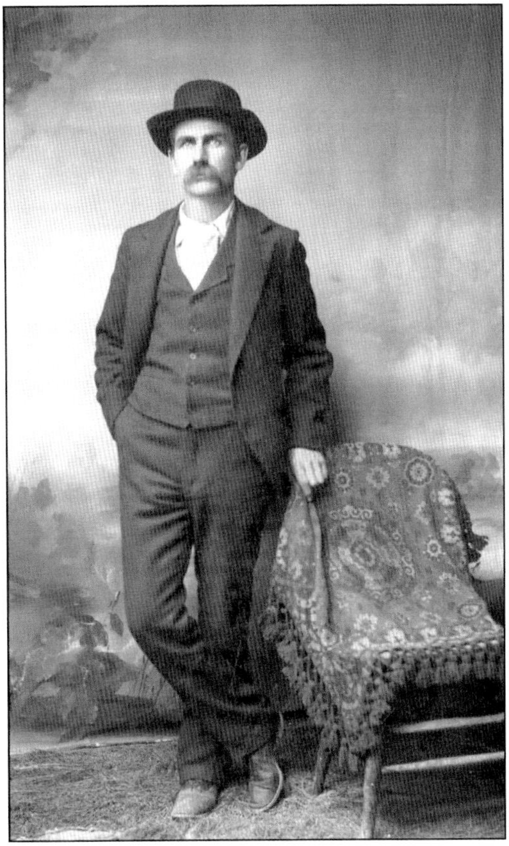

TAYLOR BROTHERS. David Taylor (below) and his brother Donald (left) came from Prince Edward Island to Hayden in July 1883. They arrived in Northwestern Colorado with fellow Canadians Albert Ryan and Archie McLachlan. Many times during the homestead period, friends would relocate together. All four men went on to Craig; the Taylor brothers later built cabins off Mack Lane near the present-day Woodbury soccer fields.

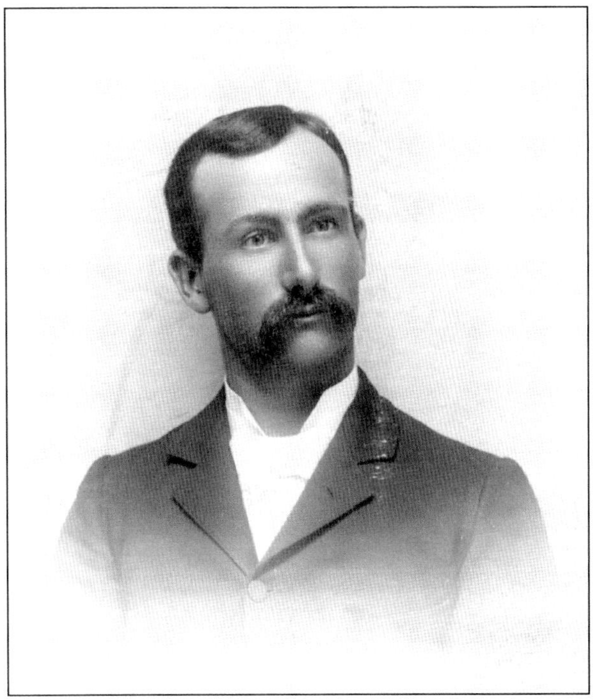

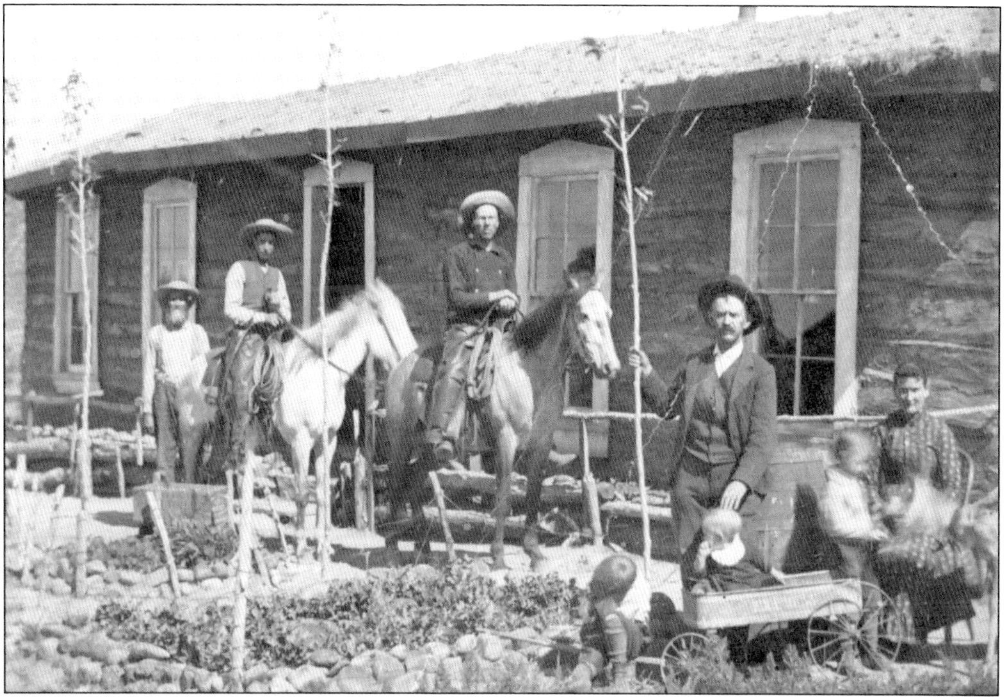

McKay Family, c. 1888. Joseph McKay and his wife Mary (right) pose with their young sons George, Earl, and Ben in front of their newly landscaped homestead cabin. Friends Francis Haughey, Oliver Haughey, and Benn McTonn are on the left. The original house, which was located just east of the present-day Craig Campground, still stands though it is now contained within new additions, and is no longer visible.

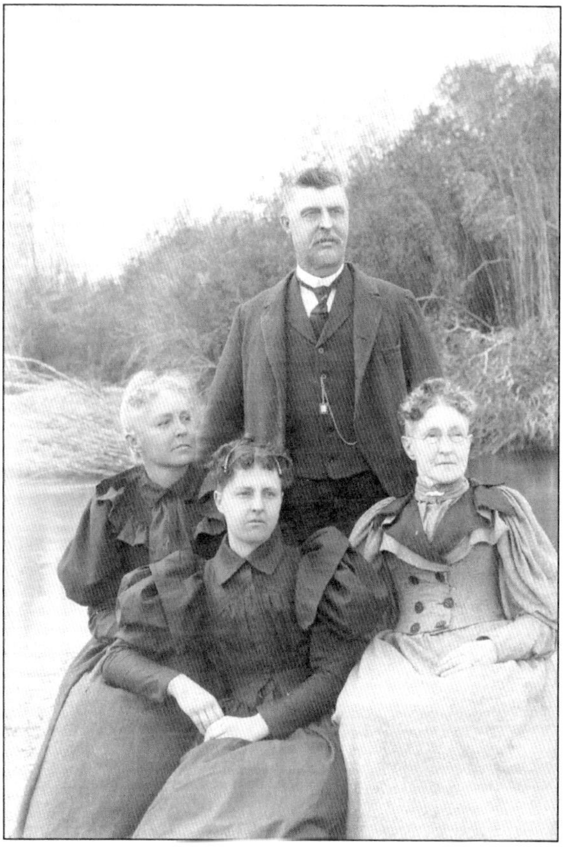

Tilton Family, c. 1897. Early homesteader Clark Tilton is pictured with his wife, Lydia, daughter Mary Blanche, and Lydia's visiting sister Margaret Hughes (right). The Tiltons came to the valley in 1883; their property is now part of the Yampa Valley Golf Course. Mary Blanche went on to homestead at South View, south of Craig, and also taught in several area schools. (Photograph by D.W. Diamond.)

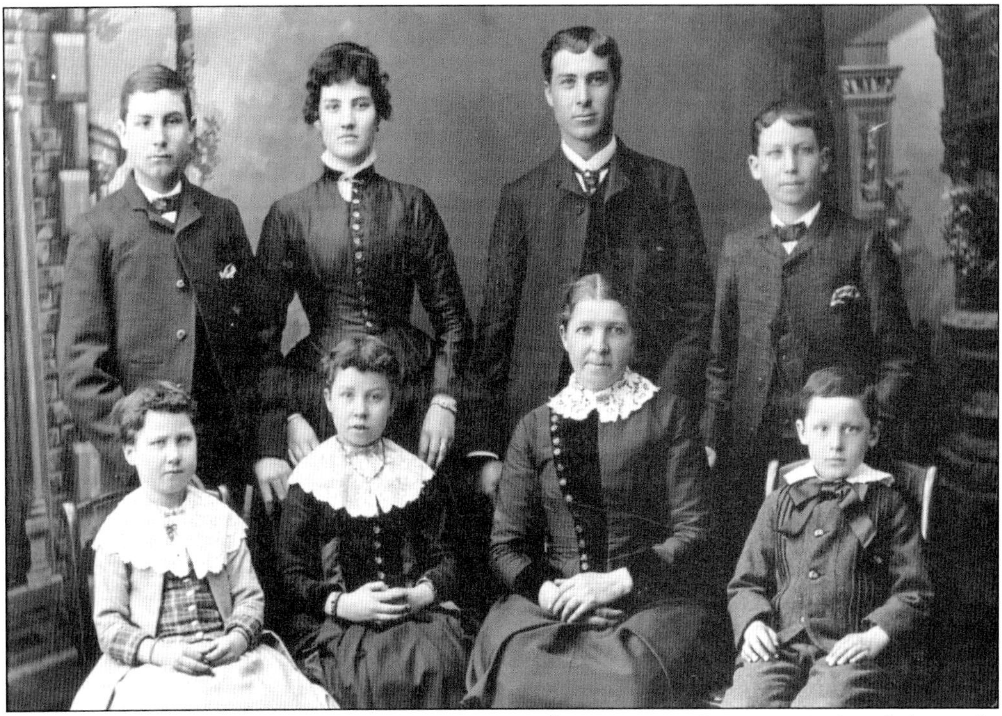

Francis Haughey Family, c. 1885. Francis Haughey came to Craig in 1886 with his wife, Isabella, and children and took up a homestead at the east end of town. He left after a few years and headed to the silver mines in Utah, leaving his wife and children to manage on their own for many years. In this photograph, taken prior to their move to Colorado, Isabella poses with her large family, but Francis is missing from the scene. Most of the Haughey children went on to become very active members of the Craig community and were instrumental in its growth. Many Haughey descendants still live in the valley today.

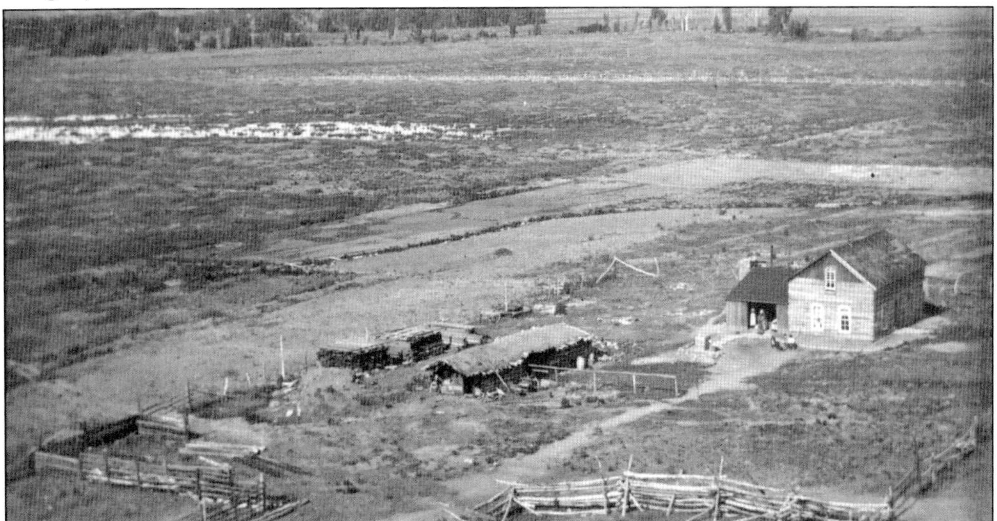

F.M. Haughey Homestead, c. 1890. If the house was still standing today, it would be located near 1424 East Victory Way. The photograph was taken from the cemetery hill looking to the southwest. The closest line of trees in the background marks Fortification Creek.

The Ranney Family. This photograph was taken before five of the siblings moved from Michigan to Northwest Colorado in 1883. From left to right are (first row, seated) Edwin, Marcia, and Alvor; (second row, standing) Charles, Frank, and Cora. Though Marcia moved only as far west as the Denver area, she made frequent visits to her siblings in the Craig area.

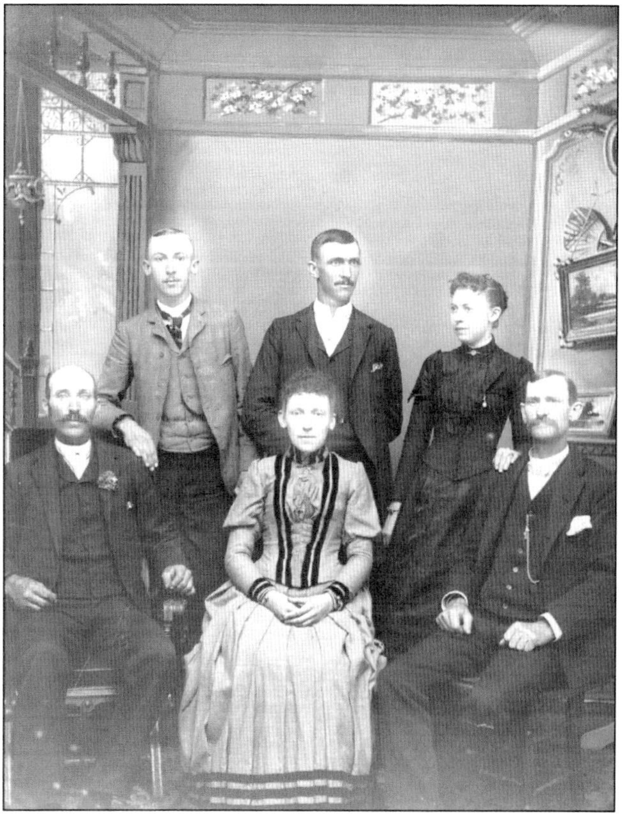

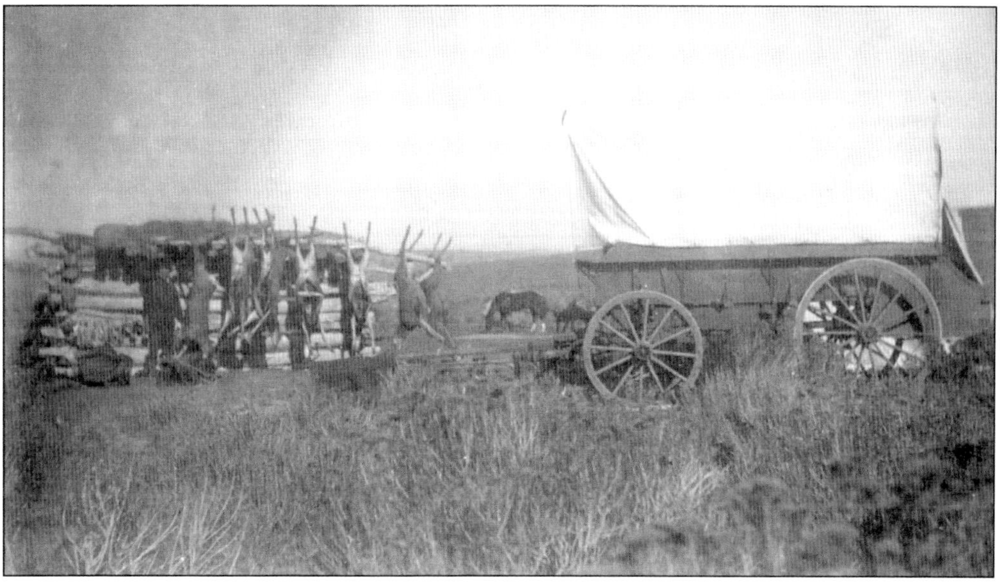

Alvor Ranney Homestead, c. 1889. Deer hang outside Alvor Ranney's cabin, which was near today's southeast corner of West Victory Way and Ranney Street. Alvor came from Michigan with his brothers in November 1883 and built a cabin the following spring on his land. Part of his homestead was purchased in 1889 by the Craig Land and Mercantile Company, which started the Craig townsite.

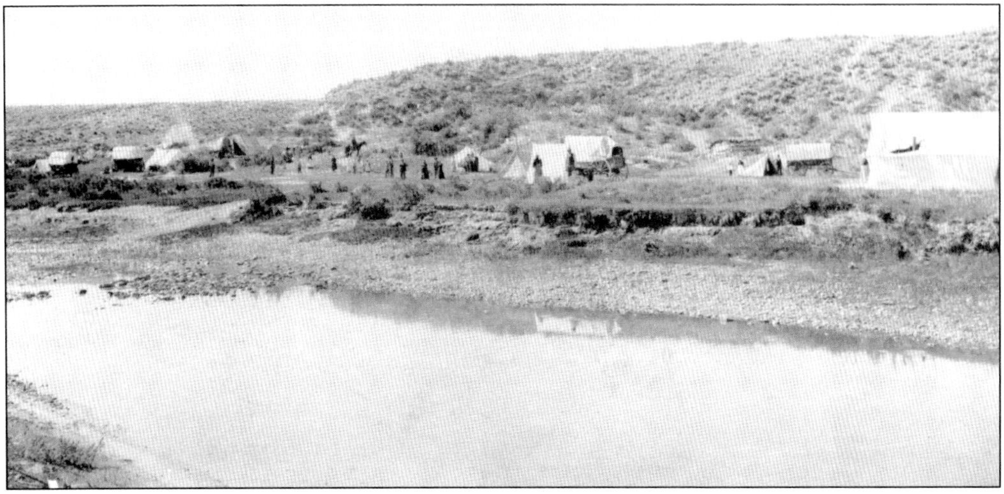

Juniper Hot Springs, c. 1892. These once-popular hot springs are located near Lay. Having heard about the springs from Ute natives, many of the early settlers would travel west to soak in the hot springs and socialize. The small log structure, seen at right center, sat over the springs. The tent at the far right with a flap in the roof was most likely owned by the photographer. This type of tent was often used by photographers to let extra light in so the tent could be used as a studio. The hot springs remained a popular local destination activity into the 1980s and are still available for use today. (Courtesy Hayden Heritage Center.)

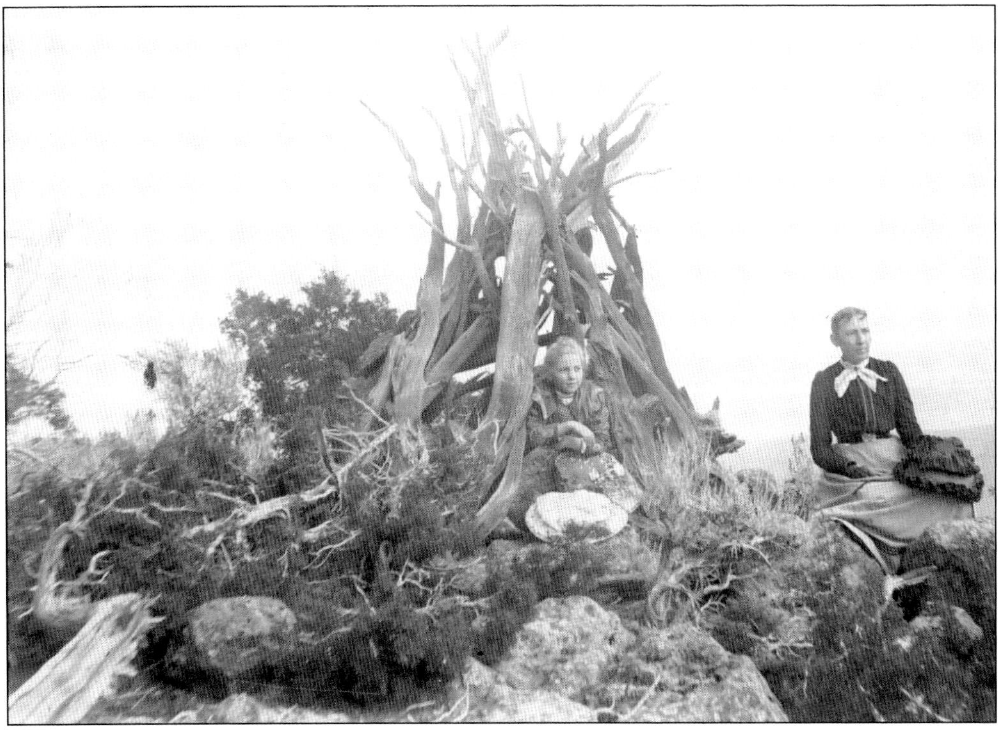

Wickiup on Cedar Mountain. Marcia (Ranney) Smith and her daughter Eva pose by a wickiup on Cedar Mountain during a visit to Craig to see Marcia's Ranney siblings around 1890. A wickiup was a Ute form of tipi, used for temporary shelter. Hiking Cedar Mountain was a favorite pastime with the early setters.

Two

Laying the Foundation

By 1888, the northwest corner of Colorado was ripe for settlement. Early homesteading families proved that the harsh climate was a surmountable obstacle and that a good living could indeed be wrested from the land in the Yampa River Basin. The river valley below the imposing grandeur of the mountains offered a place for new communities to flourish in this remote region of the state, and that notion was not lost on a group of enterprising businessmen.

William H. Tucker, having heard rumors of a pending rail connection between Denver and Salt Lake City, came in the spring of 1889. He and his brother-in-law W.F. Teagarden realized that the area around the Yampa River had potential as a townsite. Tucker returned to Denver to obtain financial backing and contracted with several men to form the Craig Land and Mercantile Company. They chose the name Craig in honor of the project's chief backers, William B. and Alexander C. Craig.

The men purchased 160 acres of land from Alvor Ranney and William H. Rose and 160 acres of state land. They erected a building to house the Craig Land and Mercantile Company, which doubled as a land sales office and general store. It also was the site of the post office for the frontier hamlet. The entire townsite was contained within the boundaries of today's First and Ninth Streets and Rose and Ranney Streets. Within two years, the town would boast a church, a school, one saloon, a hotel, two mercantile establishments, and a newspaper.

By nature, settlers and homesteaders are not a timid lot, and those coming into the area brought with them the necessary enthusiasm and energy for the work they faced. By the 1900 census, the burgeoning village had a population of 133, and with the promise of the anticipated railroad, more solid growth seemed inevitable. The town was on its way to a secure footing.

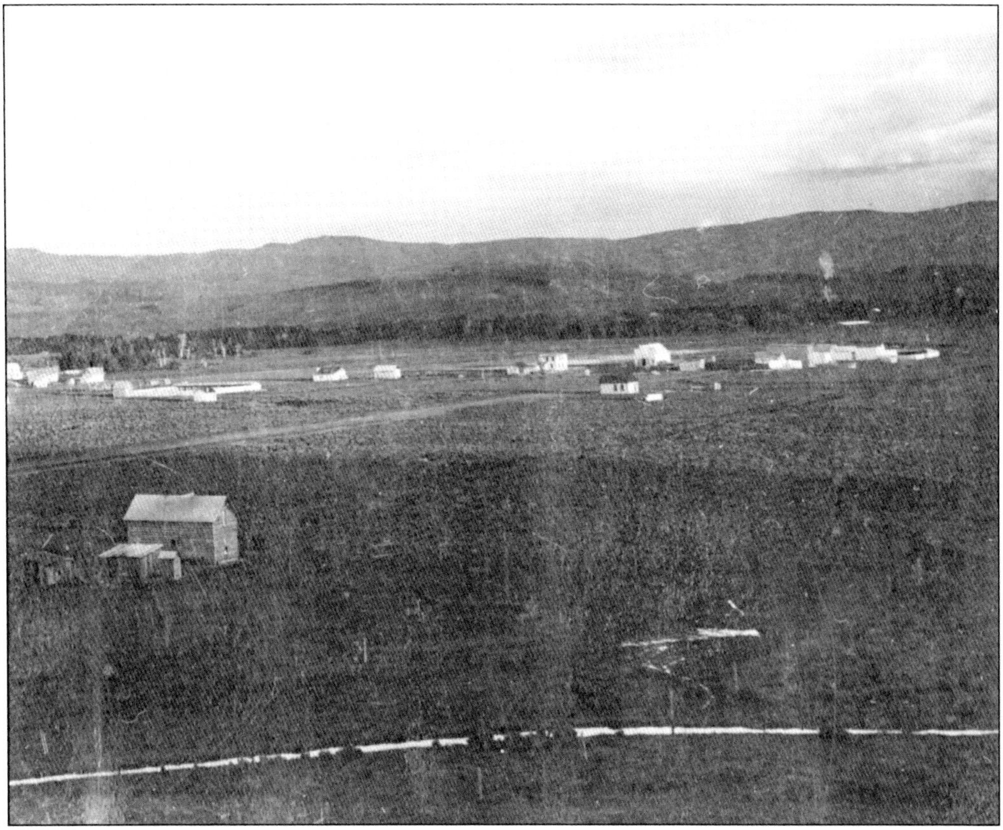

Craig, Summer of 1891. This is the earliest known view of Craig. It looks southeast over the valley from the Sandrocks.

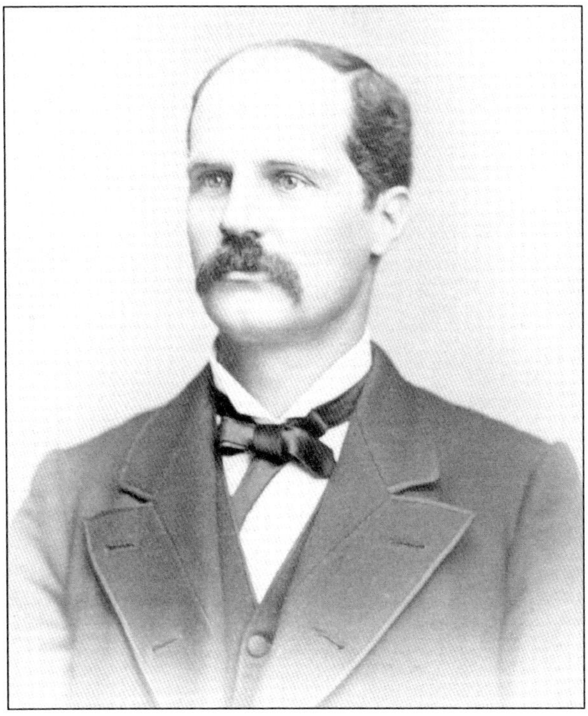

William Bayard Craig, c. 1890. There were two men with the last name of Craig, apparently unrelated, who were in the land company that founded the town of Craig in 1889. One of them, Denver minister Wm. Bayard Craig, was the major contributor to the company and the townsite was given his name.

EARLY LAND DEVELOPERS

A Mathew Johnson eventually bought by Elmer Mack. Mack sold land west of Hwy. 13 789 to Pine Ridge Development 1887

B David Taylor eventually developed in Riverview. 1886

C Fred S. Elliot sold land north of river to John Mack, now owned by Elmer Mack. 1887

D William F. Finley

E Albert Ryan eventually sold to Woodbury's. 1892

F Woodbury sold to development Corp.

G Donald Taylor 1886

H John Mack now owned by son, Elmer. 1890

J William C. Barkley bought from State, sold to Robert Green. 1891

K Frank Ranney 1890

L Archibald McLachlan subdivided in recent years by Wingos. 1891

M Bought from State Land Board by Craig Land and Mercantile Co.

N Alvor Ranney sold to Whyte to Taylors

P Alvor Ranney sold to Craig Land and Mercantile Co.

Q Alvor Ranney sold to Rose to Craig Land and Mercantile Co.

R Myrtle Bryan VanDorn mother of Bob and Perry

S Jacob Harding an early attorney in Craig

T William H. Rose

U Harvey Shaukland

V Byron B. Cooper

W Francis M. Haughey

CRAIG MAP AND MAP KEY. The map key references early homesteads and their locations in the current Craig area.

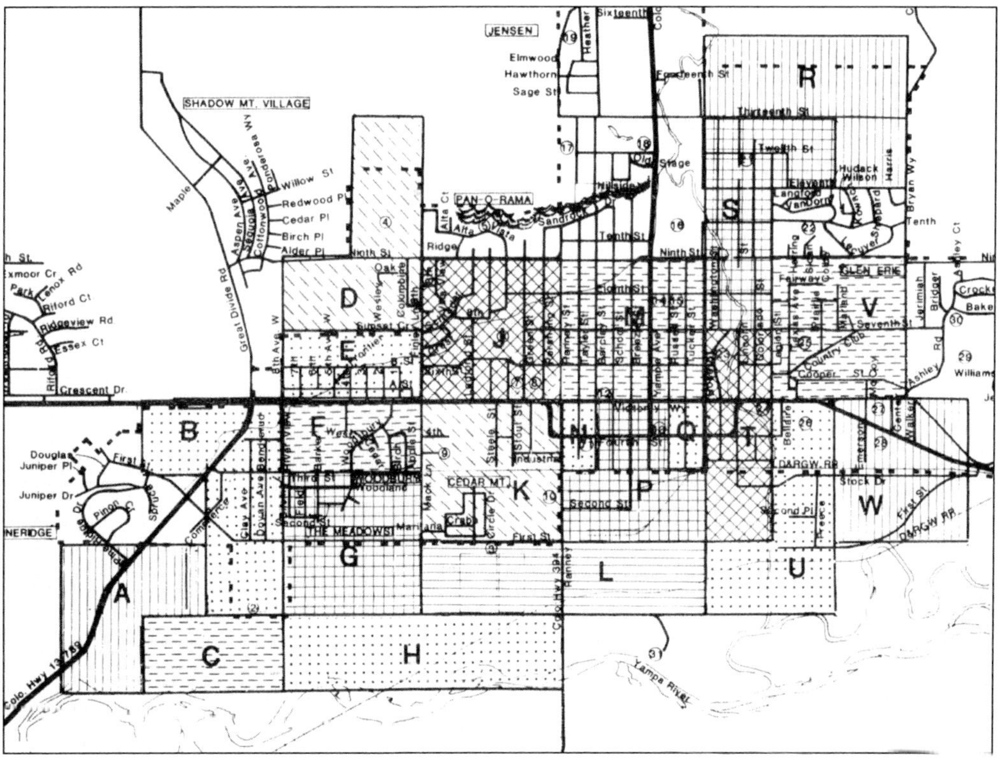

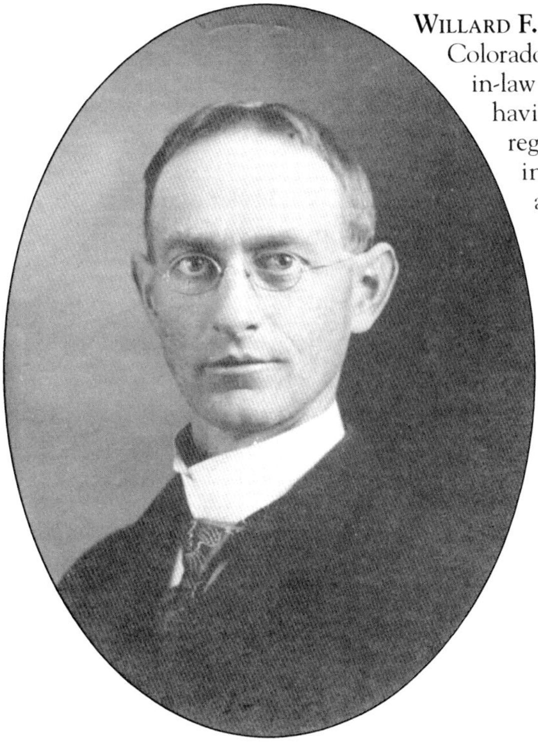

WILLARD F. TEAGARDEN, C. 1900. Teagarden moved to Colorado in 1887. Two years later, he and his brother-in-law W.H. Tucker came to Northwest Colorado, having heard of a possible rail route through the region. Deciding that development was feasible in the area, they organized the Craig Land and Mercantile Company, and with the help of financial backers, laid out the townsite in 1889.

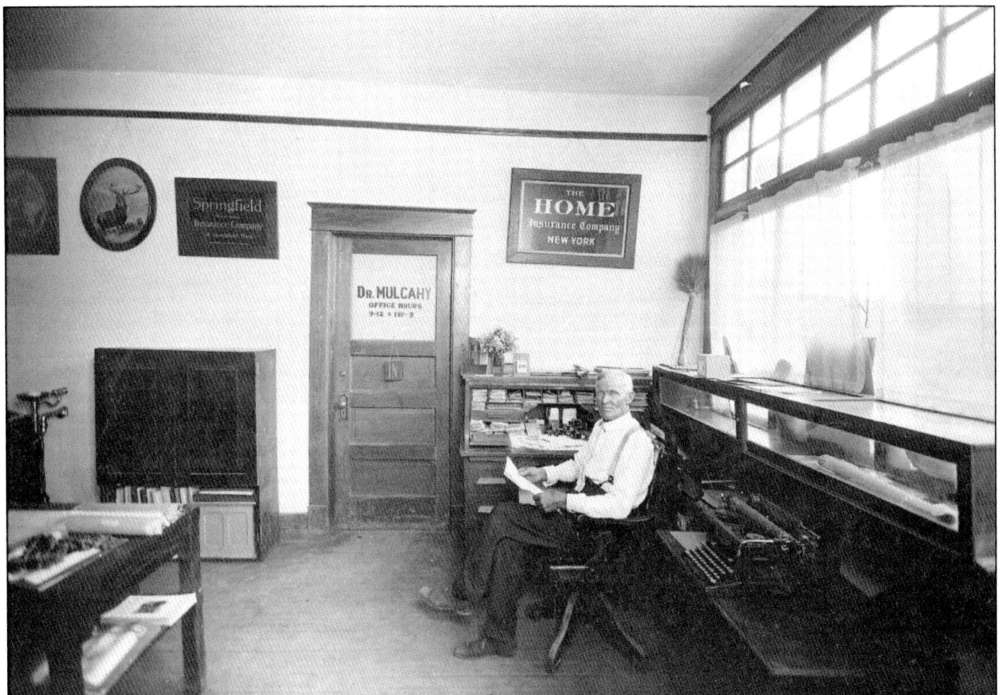

WILLIAM H. TUCKER. Tucker and Teagarden platted the Craig townsite in 1889. He is shown here in his insurance office, many years later, which was located at 530 Yampa Avenue, home to present-day Kester's Jewelry.

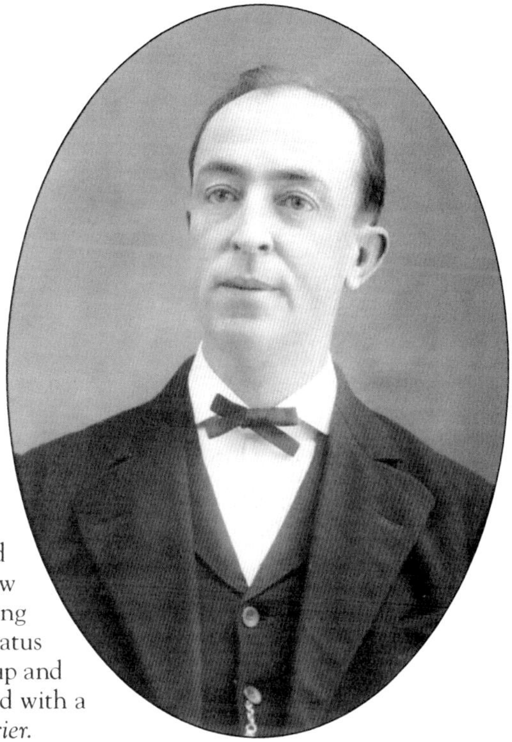

CLARENCE AND GRACE BRONAUGH. Bronaugh arrived in Craig behind a four-horse team pulling a wagonload of printing equipment on March 12, 1891. This brash young editor set up his press on the boardwalk of Yampa Avenue and published his first issue of the *Pantagraph* the following day. In 1893, Bronaugh went to Illinois and returned with his new bride, Grace. The Bronaughs left in November 1894, and until his last day in Craig, Clarence remained an avid booster of the growing frontier town. The new owners of the paper were unable to keep it going past March 1895. There was a several-month hiatus before this voice for the little community was up and running again. In July, publication was resumed with a new publisher and a new name, the *Craig Courier.*

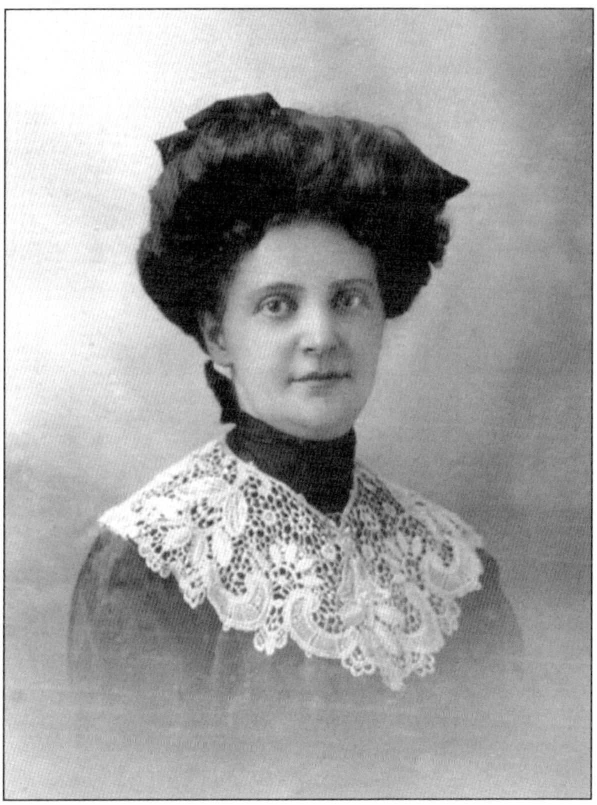

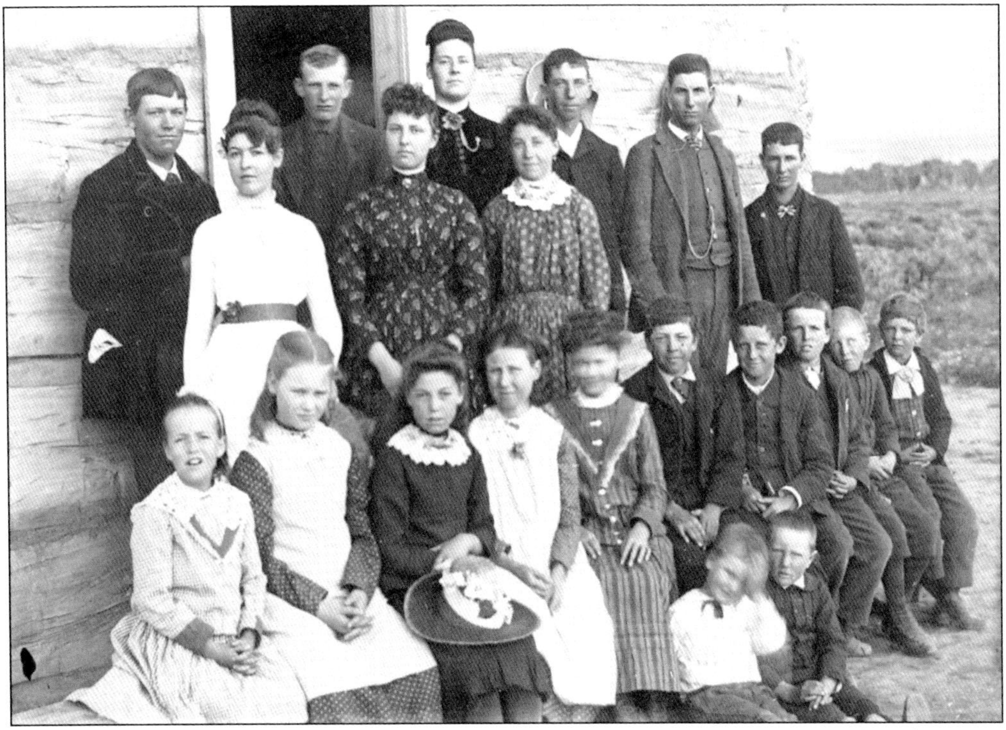

Craig School, 1890. This first school was built around 1887 and stood near Fourth and Ranney Streets. It served the homesteading families until a new two-story building was constructed in 1892. Three students identified in this photograph are Blanche Tilton, Ossa Haughey, and Maud Haughey. Two of the Haughey brothers are also pictured. Teacher Rose Johnson is in the center of the back row.

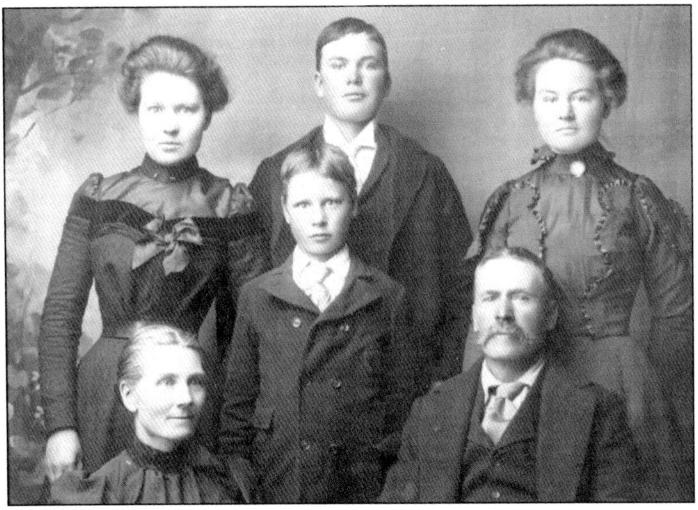

George Givens Family, c. 1898. Born in Scotland, George Givens immigrated to the United States and came west sometime before 1880. By 1885, George was listed in a Colorado census as a resident of the village of Yampa. Though his homestead was outside of town, he built his house (the first one in Craig) in 1889, on the east side of the 500 block of Russell Street, so his children could go to school.

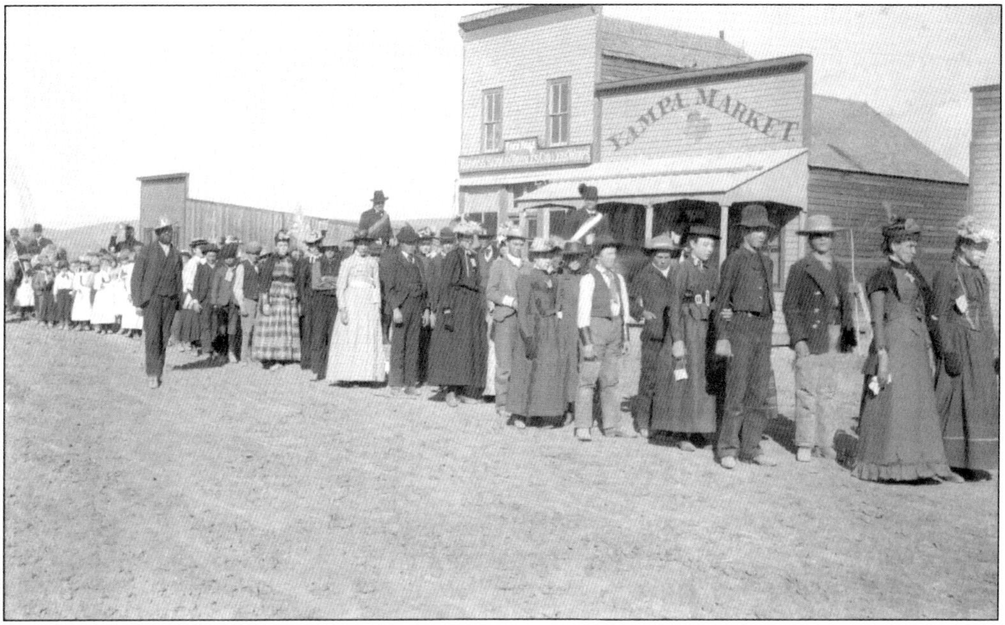

COLUMBUS DAY PARADE, 1892. By the autumn of 1892, the town boasted enough residents to hold a parade to celebrate Columbus Day. These schoolchildren participate in the holiday parade as it heads south in the 500 block of Yampa Avenue. (Photograph by W.O. Luke of Leadville.)

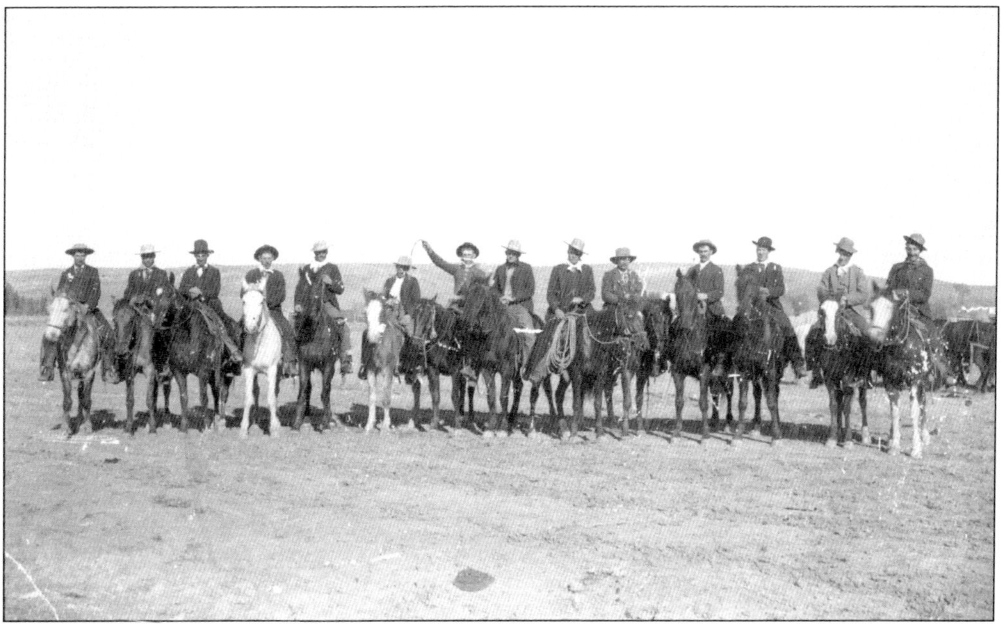

COWBOYS IN THE PARADE, 1892. Local cowboys line up to take part in the parade on Yampa Avenue. They are, from left to right, Charles Taylor, Oliver Haughey, Sid Pughe, John Slates, Ed Sizer, Charles Frederick, James Adair, Thomas Clark, Alex Walker, Ed Sax, Mark King, Walter Crowell, John McKeever, and Joseph Haubrich.

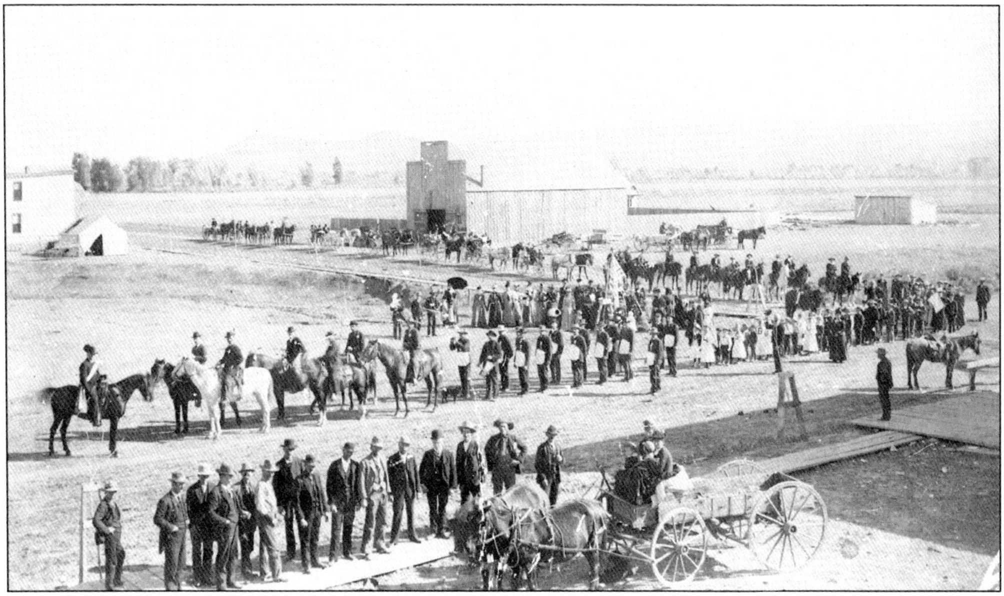

Parade Participants, 1892. With so many participants in the Columbus Day parade, it is a wonder that there were any people left to view the procession. Local law enforcement officers, Masonic lodge members, schoolchildren, cowboys, and residents showing off their horse teams and rigs lined up along Main Street and Yampa Avenue to take part in the event.

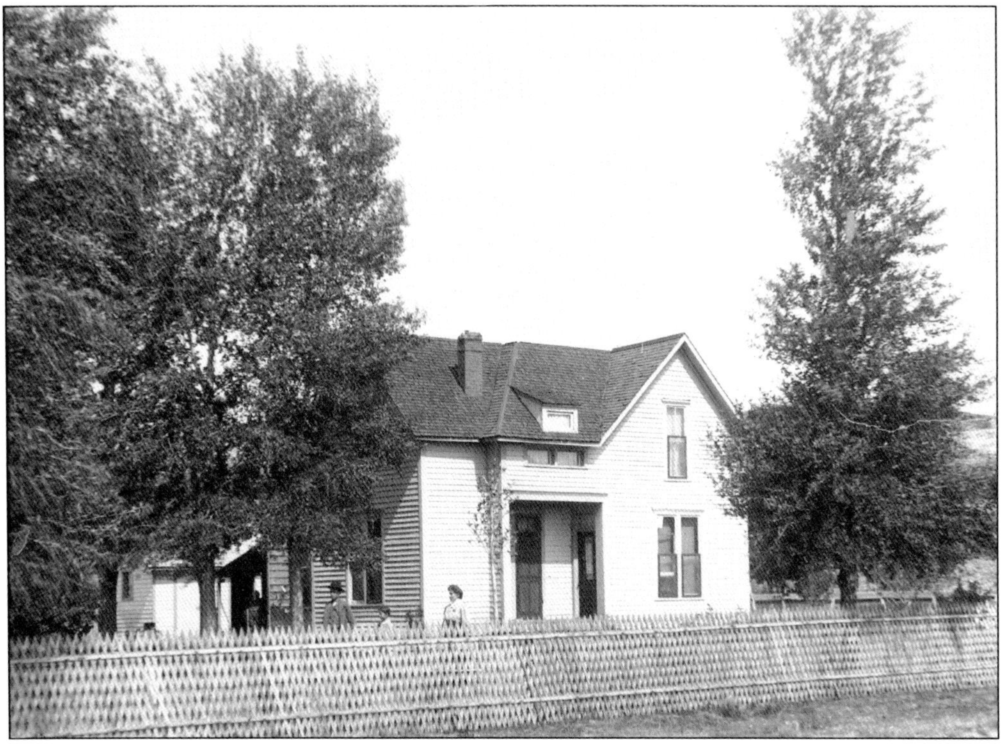

William H. Tucker Home. Tucker, who was instrumental in the founding of the Craig townsite, built this house about 1890 at 806 Yampa Avenue. The home is still a landmark in Craig.

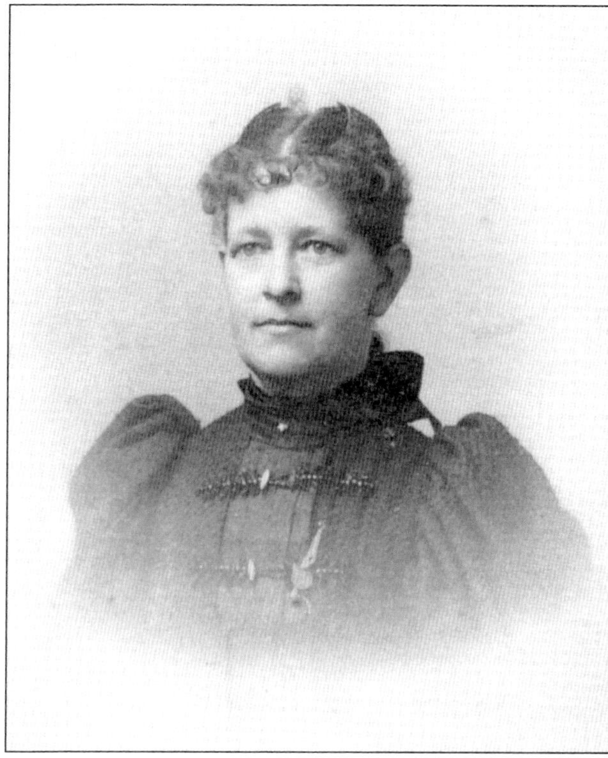

Flora Teagarden Tucker and Rosella Teagarden Breeze. The two Teagarden sisters were born in West Virginia just before the Civil War. Flora (right) married William Tucker in Colorado and moved with him to Craig in 1889. Rosella (below) moved to Colorado for her health in 1888 and made her home with Flora and William until she married early area settler Lemuel Breeze in 1891. The sisters were noted for their kindness, striving to help newcomers settle into the close-knit frontier community of Craig. Willard Teagarden, brother of the two sisters, worked with William H. Tucker to lay out the Craig townsite in 1889. The Teagardens' half-brother Alexander Robinson also came to Craig in 1890 and was elected the town's first mayor in 1908.

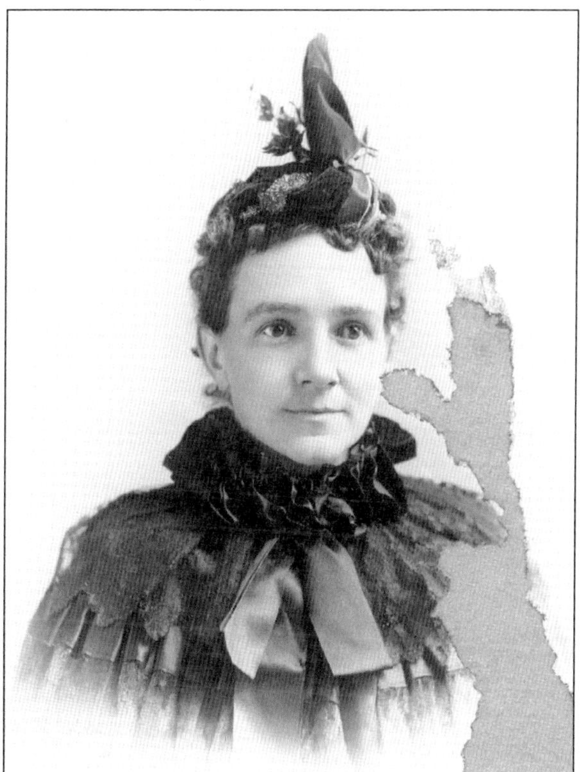

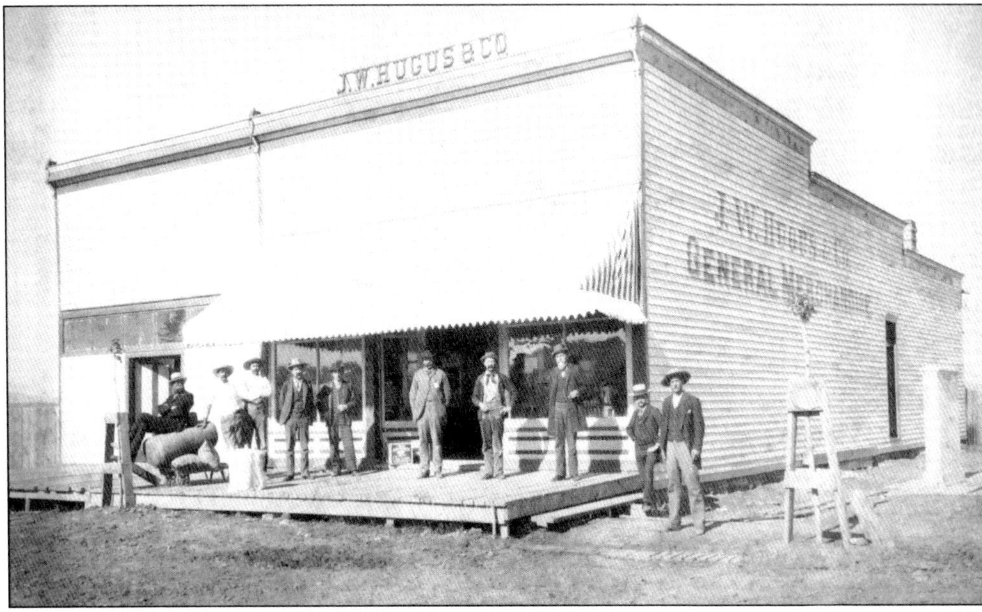

J.W. Hugus & Company, c. 1892. Located on the southwest corner of Main Street (Victory Way) and Yampa Avenue, this store was built in 1890. It was one of numerous regional J.W. Hugus stores located throughout Colorado and Wyoming. This building served the Hugus company until 1916 and continued to be used commercially until it was torn down in 1971. (Both photographs by A.G. Wallihan.)

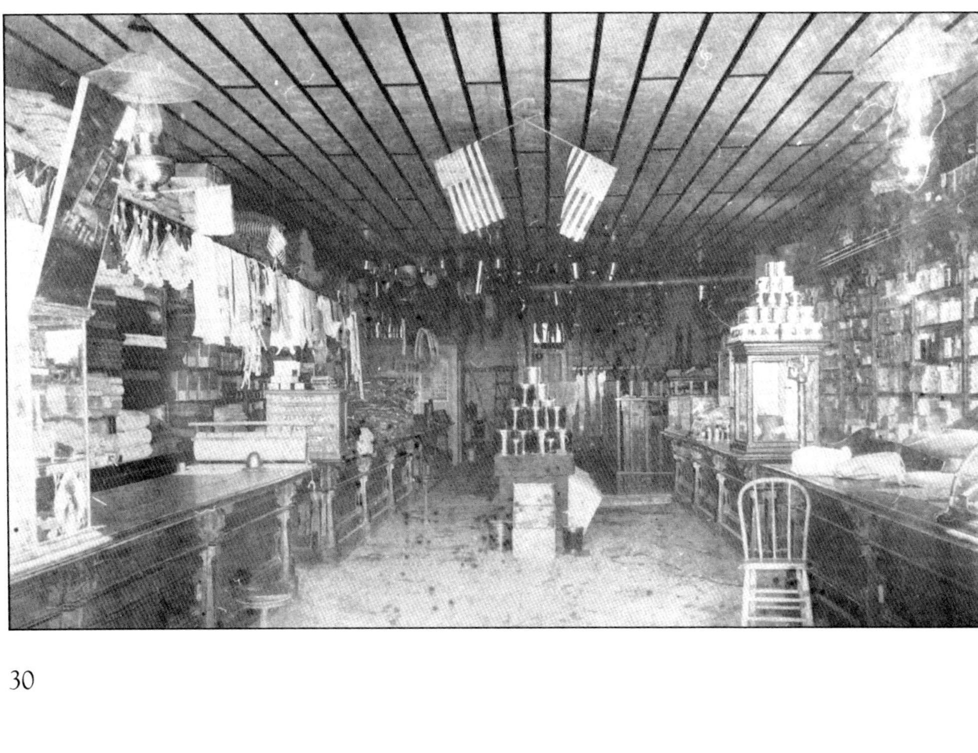

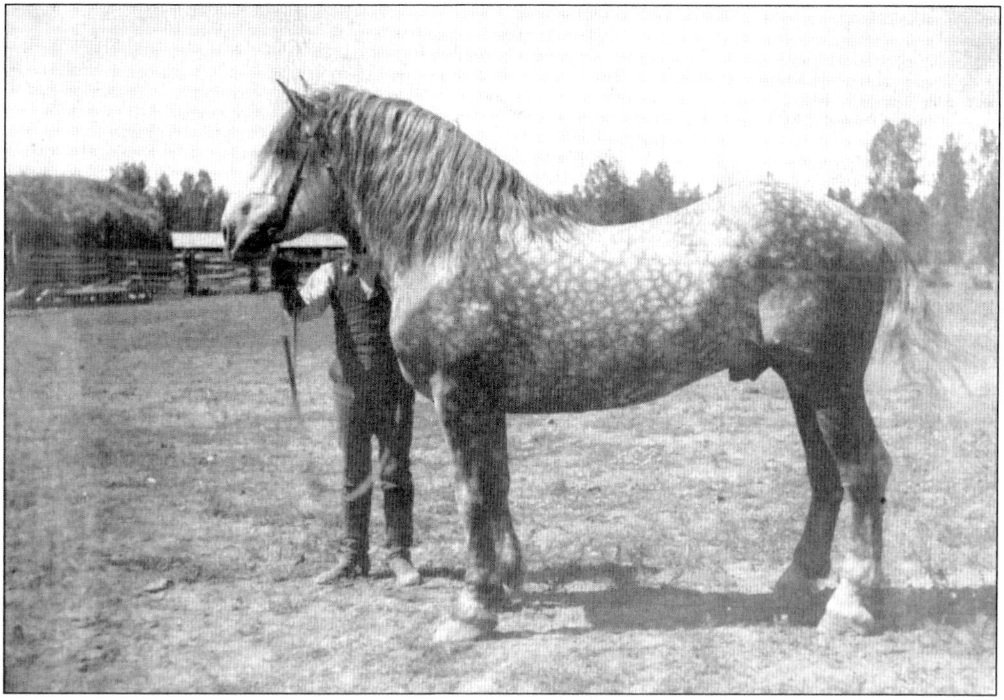

QUALITY RANCH STOCK, 1892. Early homesteaders went to great lengths to import good horses to build local herds. Brutus, a gray registered Percheon stallion, and his handler are pictured in 1892. (Photograph by A.G. Wallihan.)

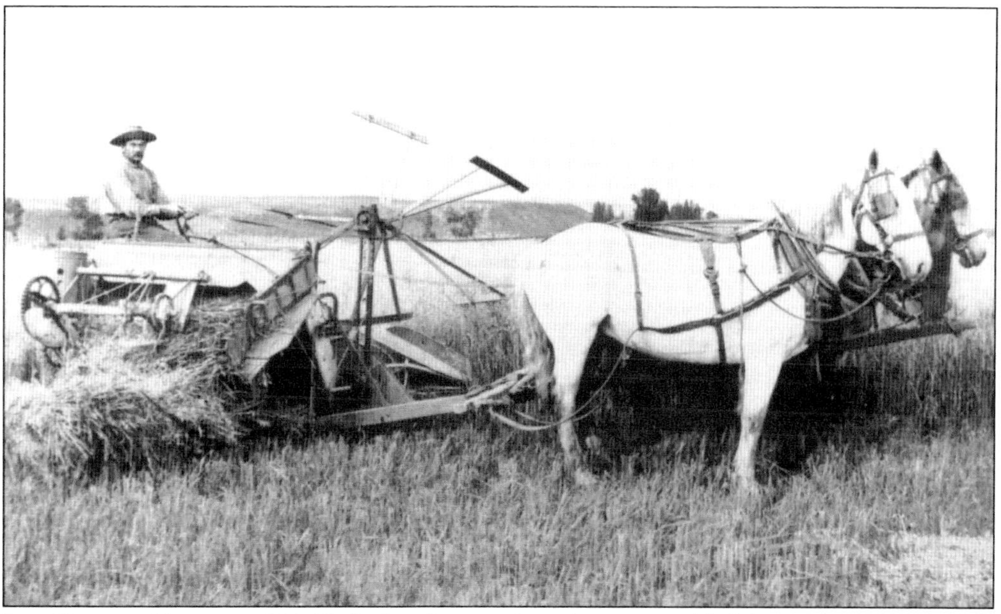

THE CUTTING EDGE, C. 1892. Early threshers, such as this one, enabled Craig's farmers to greatly increase their grain harvests. This photograph shows a harvest in process in the Yampa Valley about 1892. The greater productivity of American farmers using new agricultural innovations contributed to the 1893 recession. Within the next few years, prices for farm-produced commodities slumped due to overproduction. (Courtesy Hayden Heritage Center; photograph by A.G. Wallihan.)

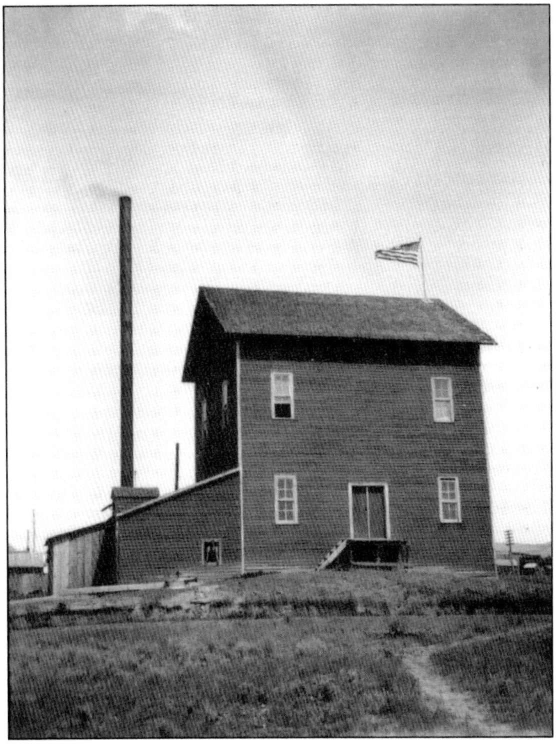

Craig Milling & Elevator Company, 1907. With increasing grain crops harvested in the area, a mill became a necessity for the young agricultural community. This mill was built in 1892 at the northeast corner of School and Sixth Streets. The mill burned in 1915. (Photograph by Robert Mark Richardson IV.)

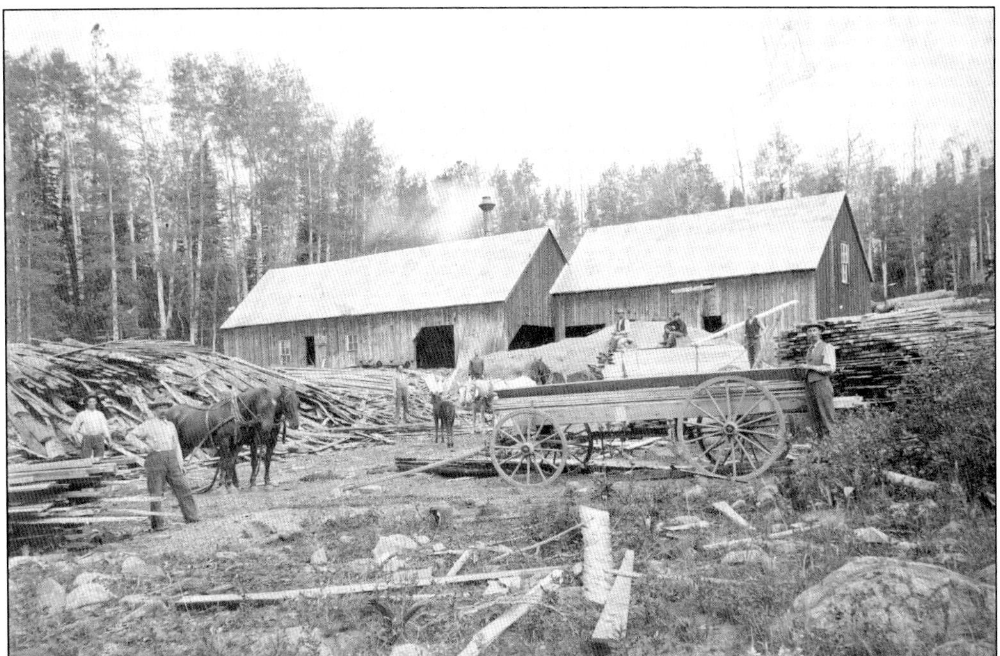

Sawmill on Black Mountain, 1896. As the young town began to develop, lumber was needed. Archie McLachlan, who had come to the area in 1883, started a sawmill operation on Black Mountain in the early 1890s. Located 25 miles north of Craig, the mill provided much of the lumber used to build the town's early structures. (Photograph by D.W. Diamond.)

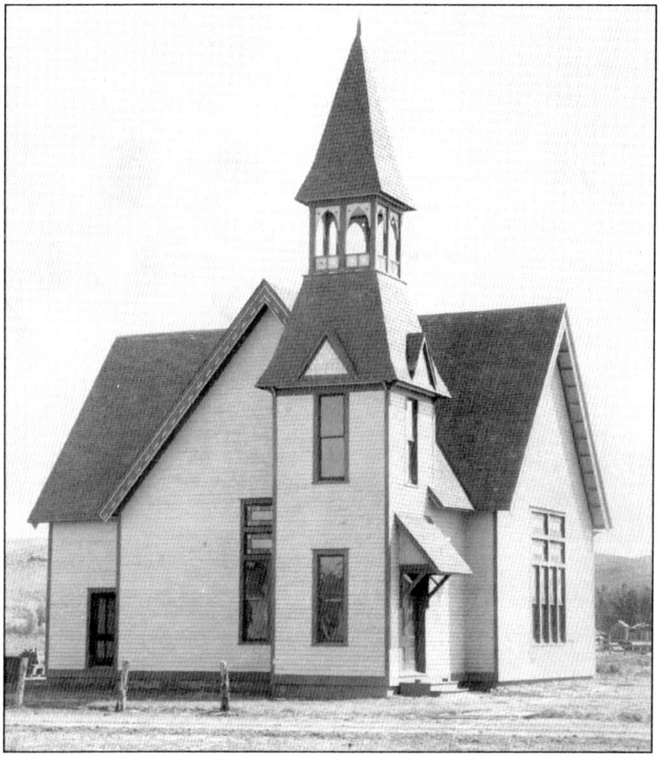

THE CHRISTIAN CHURCH, SEPTEMBER 1895. This is the first church building in Craig. Construction started in 1893 and was completed in 1894. The bell, always an important addition in a frontier community, was added in September 1898. The church burned on February 14, 1901, but was rebuilt the following year. The bell survived the fire and is still hanging in the building's belfry today. (Photograph by W.J. Johnston)

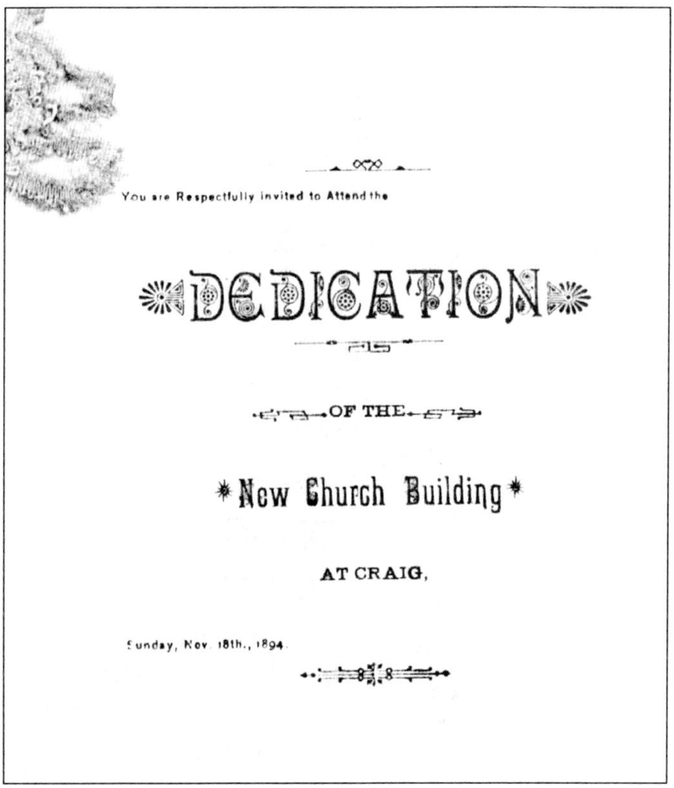

You are Respectfully invited to Attend the

DEDICATION

OF THE

New Church Building

AT CRAIG,

Sunday, Nov. 18th., 1894.

CHRISTIAN CHURCH DEDICATION. The Christian church was dedicated on Sunday, November 18, 1894.

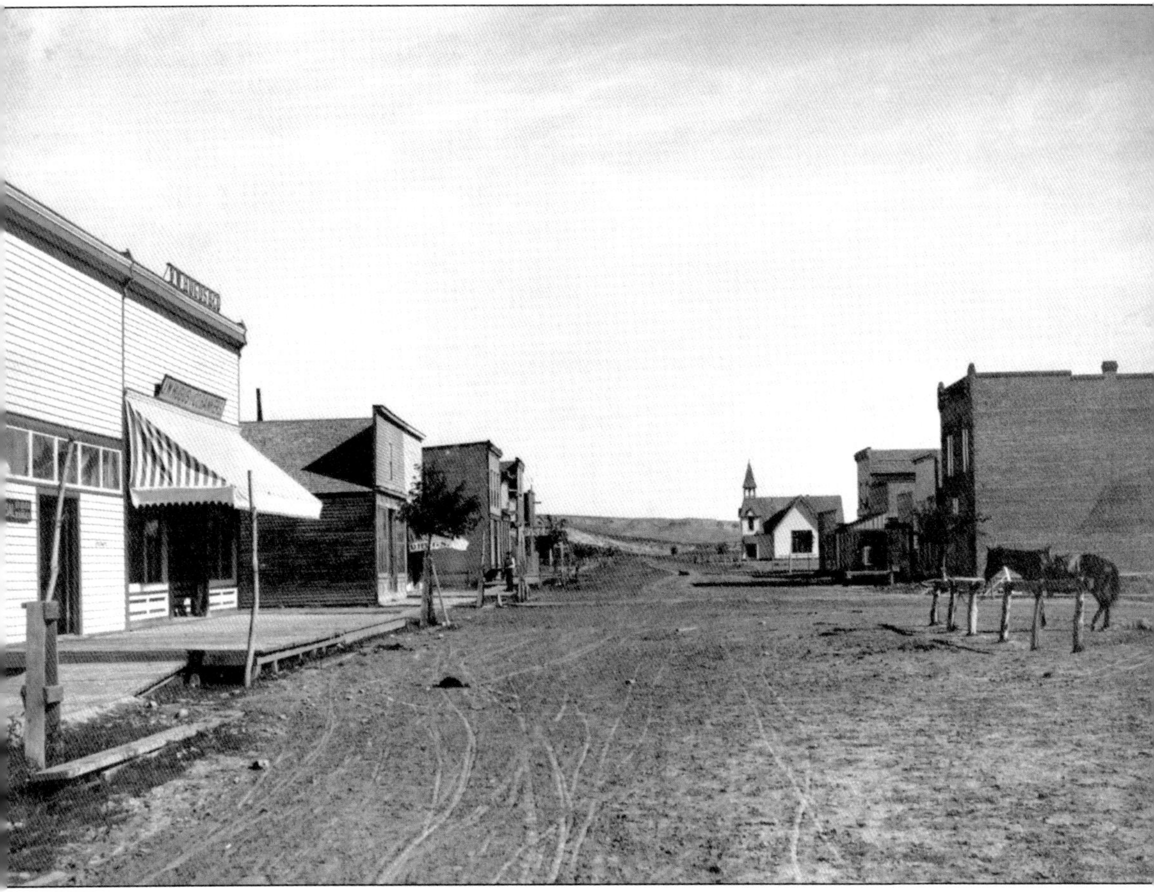

CRAIG, SEPTEMBER 1895. By 1895, Craig was developing into a "real" town, with storefronts lining both sides of Yampa Avenue. The lower end of Yampa Avenue looking north shows a building under construction (right), which is still in use today. The structure at the far left was built in 1890 and housed the J.W. Hugus & Company store. The second building on the left was constructed by the Craig Land and Mercantile Company in 1889 and housed the community's first mercantile, owned by William Tucker. The town's first church is visible at the end of the street on the right. (Photograph by W.J. Johnston)

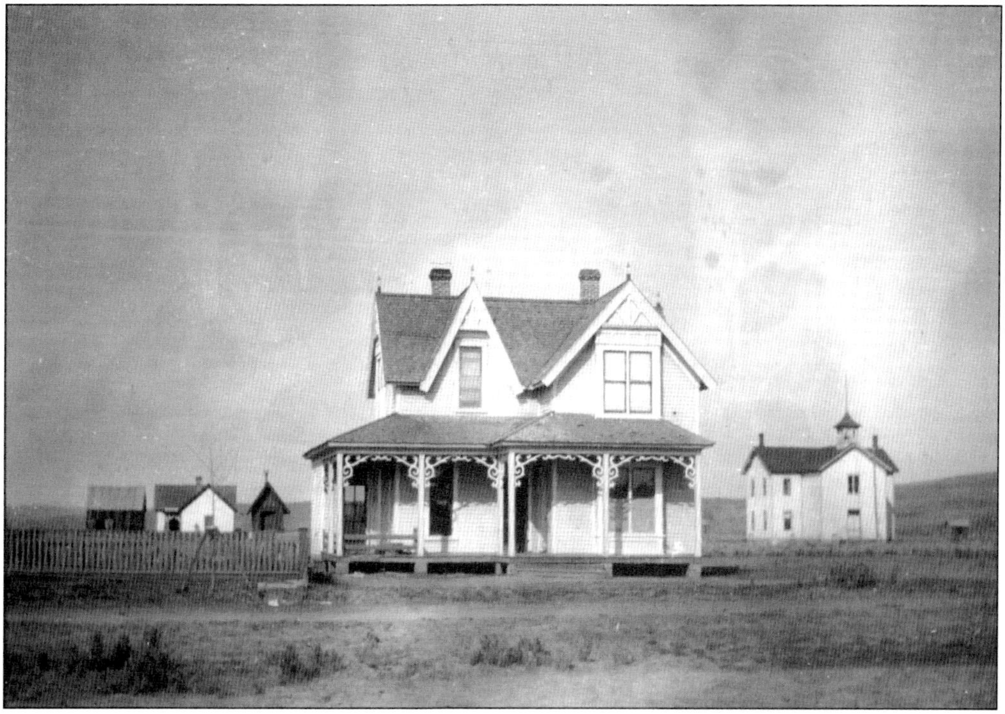

Humphrey Jones Home, c. 1897. Jones was the editor of the *Craig Courier*, the newspaper which he resurrected in 1895 from the defunct *Pantagraph* newspaper; his home at 650 Yampa Avenue is pictured here. Built in 1895 for Dr. Bennett, the house still stands. (Courtesy Denver Public Library; photograph by Abraham Fellows.)

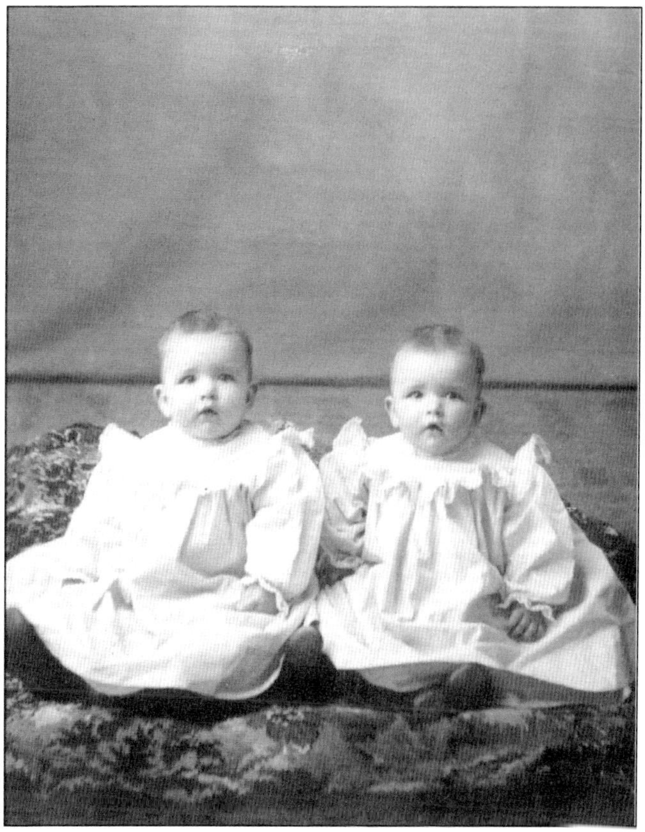

Twin Baby Girls, 1897. Ruth and Esther Jones, shown here at 10 months, were the daughters of Humphrey Jones and Annette Jane Colver Jones. The babies arrived on February 25, 1897, and were possibly the first set of twins born in the area. (Courtesy Tread of Pioneers Museum.)

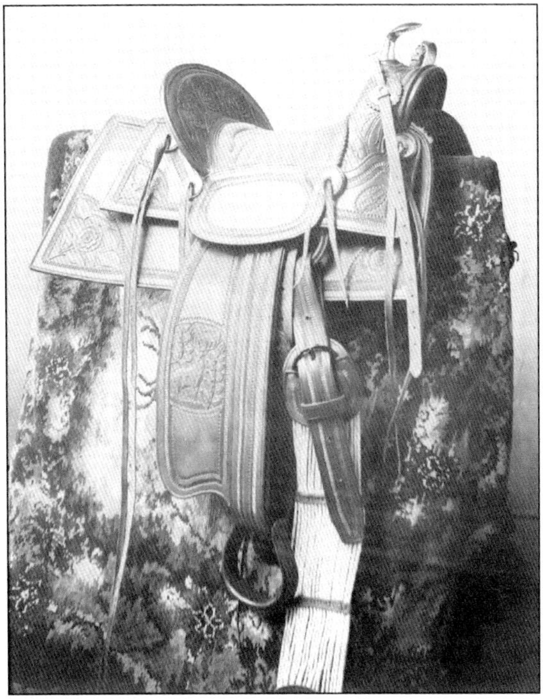

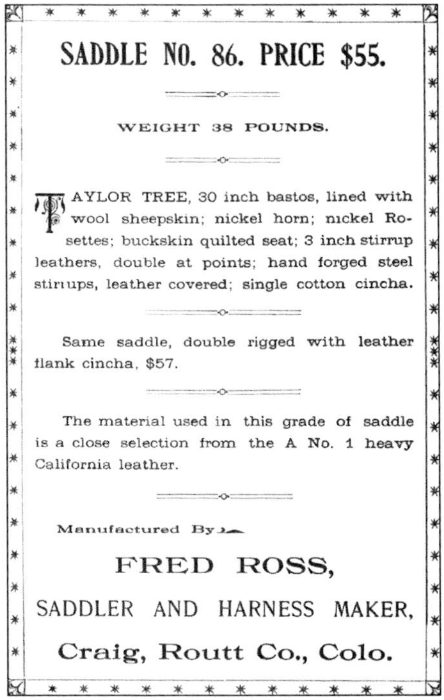

SADDLE NO. 86. PRICE $55.

WEIGHT 38 POUNDS.

TAYLOR TREE, 30 inch bastos, lined with wool sheepskin; nickel horn; nickel Rosettes; buckskin quilted seat; 3 inch stirrup leathers, double at points; hand forged steel stirrups, leather covered; single cotton cincha.

Same saddle, double rigged with leather flank cincha, $57.

The material used in this grade of saddle is a close selection from the A No. 1 heavy California leather.

Manufactured By

FRED ROSS,

SADDLER AND HARNESS MAKER,

Craig, Routt Co., Colo.

FRED ROSS SADDLE CARD, 1898. Before the arrival of the automobile, a town of any size would usually have a saddler. This promotional card was made for Fred Ross, a young German immigrant who came to Craig in 1891 and operated a saddle shop; he moved to Kentucky about 1905. Note that a saddle cost $55 at the time. (Photograph by D.W. Diamond.)

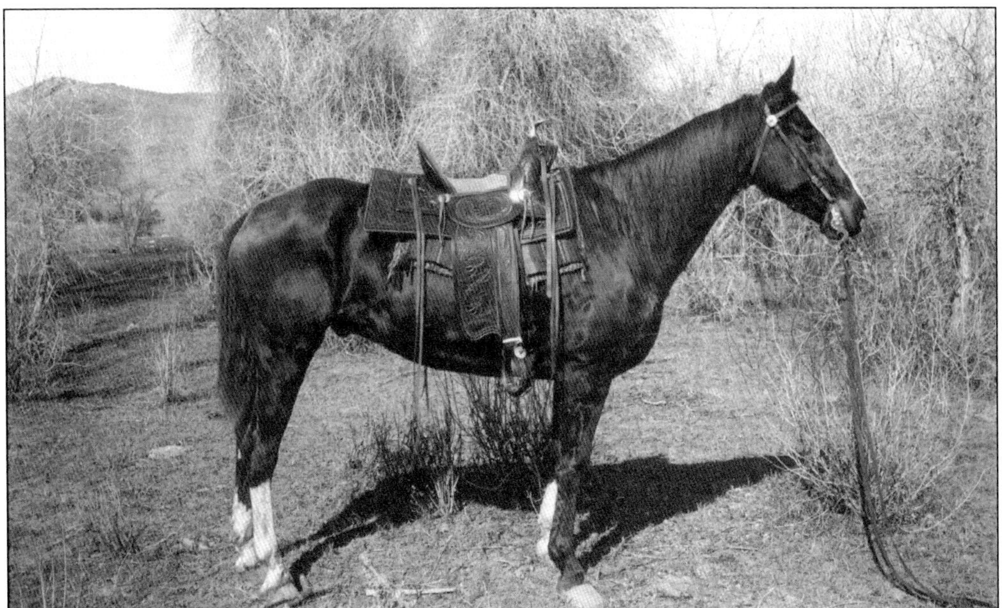

HORSE AND SADDLE, C. 1900. This horse is pictured with a saddle made by Fred Ross. Ross's saddle shop was first located at 525 Yampa. He later moved to another undetermined location in Craig. (Photograph by D.W. Diamond.)

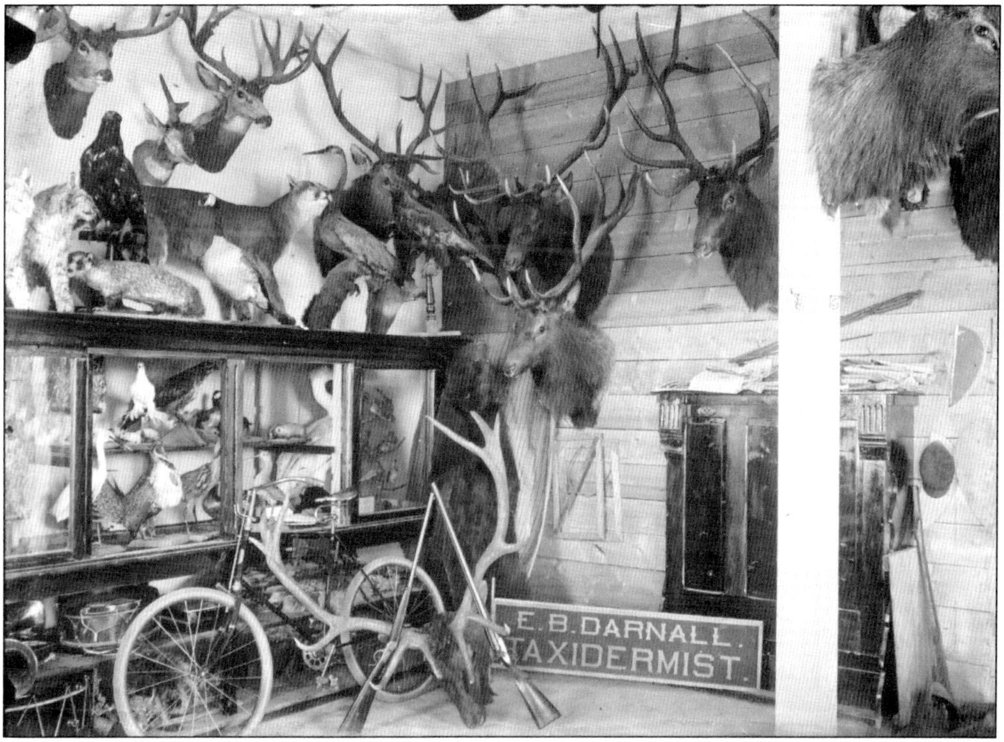

TAXIDERMY SHOP, C. 1900, AND DARNALL LETTERHEAD. Craig offered outstanding big-game hunting opportunities, and taxidermy was an essential service for the small frontier town. E.B. Darnall opened his taxidermy shop in 1894 at 509 Yampa, pictured here. (Photograph by D.W. Diamond.)

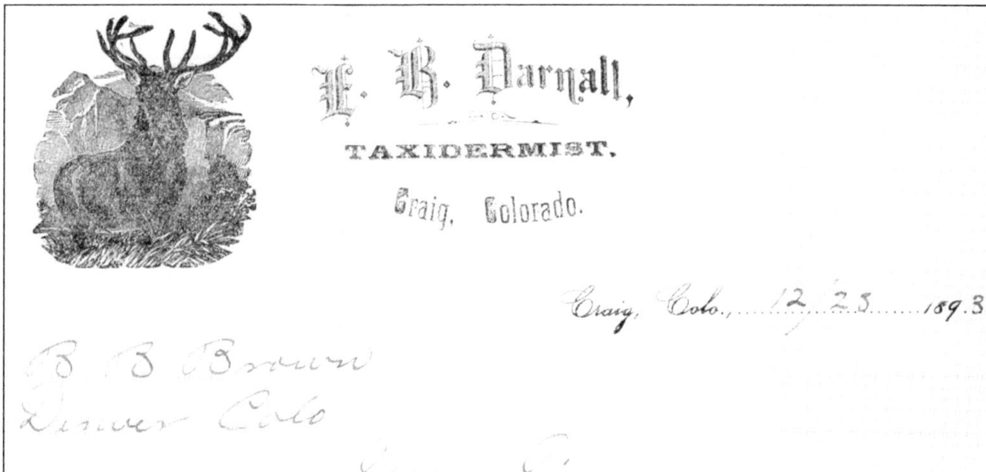

E. B. Darnall,

TAXIDERMIST.

Craig, Colorado.

Craig, Colo., 12 25 1893.

B B Brown
Denver Colo

Dear Sir

DARNALL LETTERHEAD. E.B. Darnall used this letterhead in 1893.

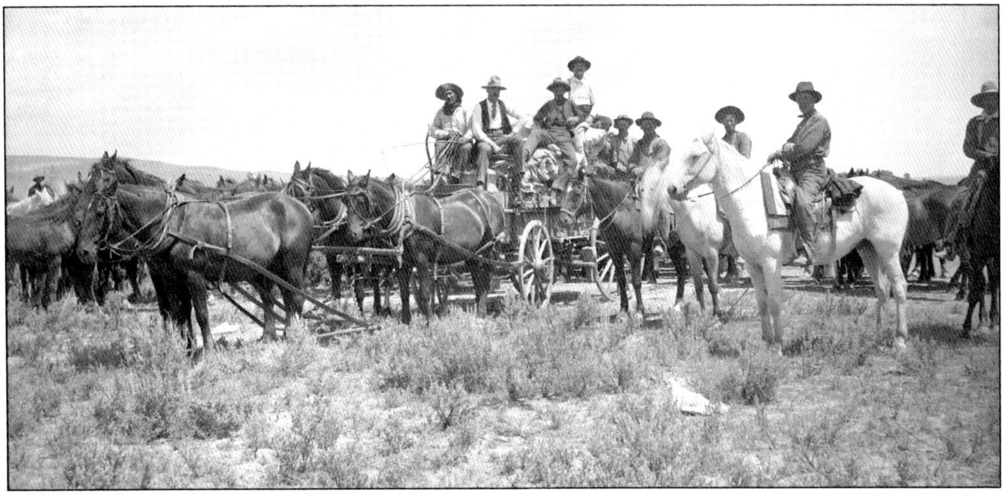

Cowboys and Mess Wagon, 1899. Cowboys are pictured with their mess wagon at Craig. The church steeple is visible behind the cowboy on the far left. The man in the white shirt and tie, seated on the wagon seat, is John Ledford. (Photograph by D.W. Diamond.)

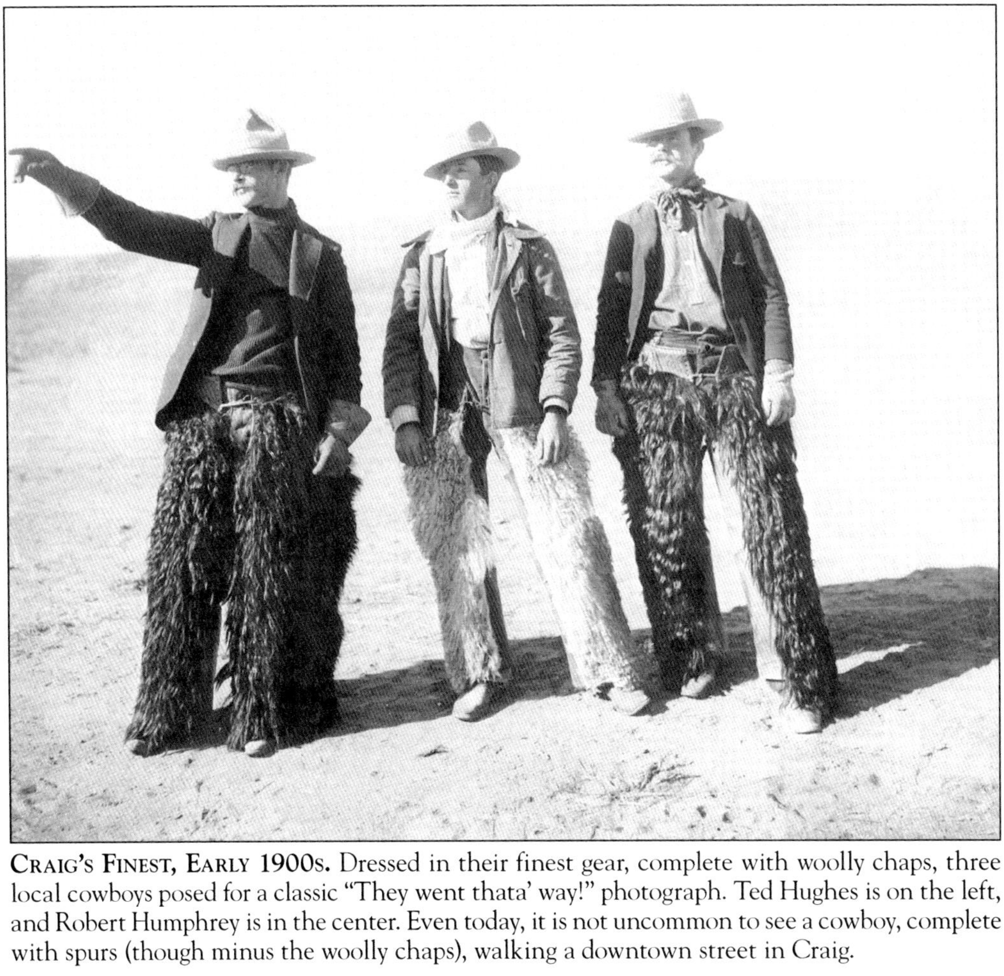

Craig's Finest, Early 1900s. Dressed in their finest gear, complete with woolly chaps, three local cowboys posed for a classic "They went thata' way!" photograph. Ted Hughes is on the left, and Robert Humphrey is in the center. Even today, it is not uncommon to see a cowboy, complete with spurs (though minus the woolly chaps), walking a downtown street in Craig.

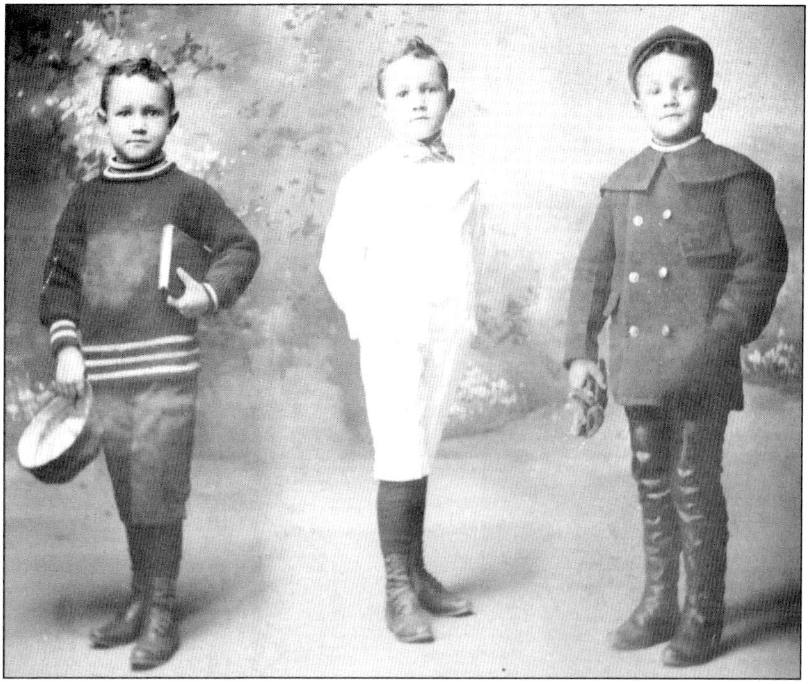

LEON BREEZE. Leon, born in 1892, was the only son of Lemuel and Rosella Breeze. This image shows three views of young Leon in different outfits. Leon's father, Lemuel, had several Breeze family relatives in the area, and Breeze Street takes its name from this family. (Photograph by D.W. Diamond)

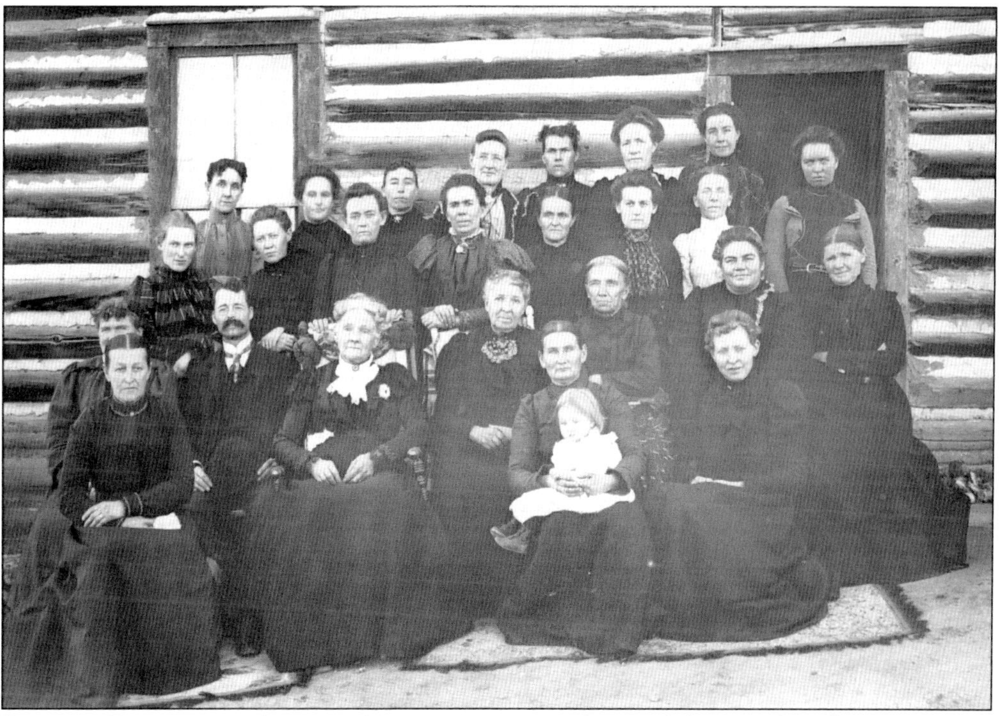

BIRTHDAY PARTY, 1901. A group of Craig area women, along with Reverend Anderson, gathered to celebrate Amanda Cooper's 70th birthday party at her home on East Main Street. Some of the identified well-wishers were Margaret Taylor, Fannie Kimberly, Alice Daniels, Ossa Cooper, Lizzie Flanagan, Jennie Earley, Ella Ledford, Ratcliff, Edith Howard, Isabella Haughey, Mary Humphrey, Cora McLachlan, and Mary McKay. (Photograph by D.W. Diamond.)

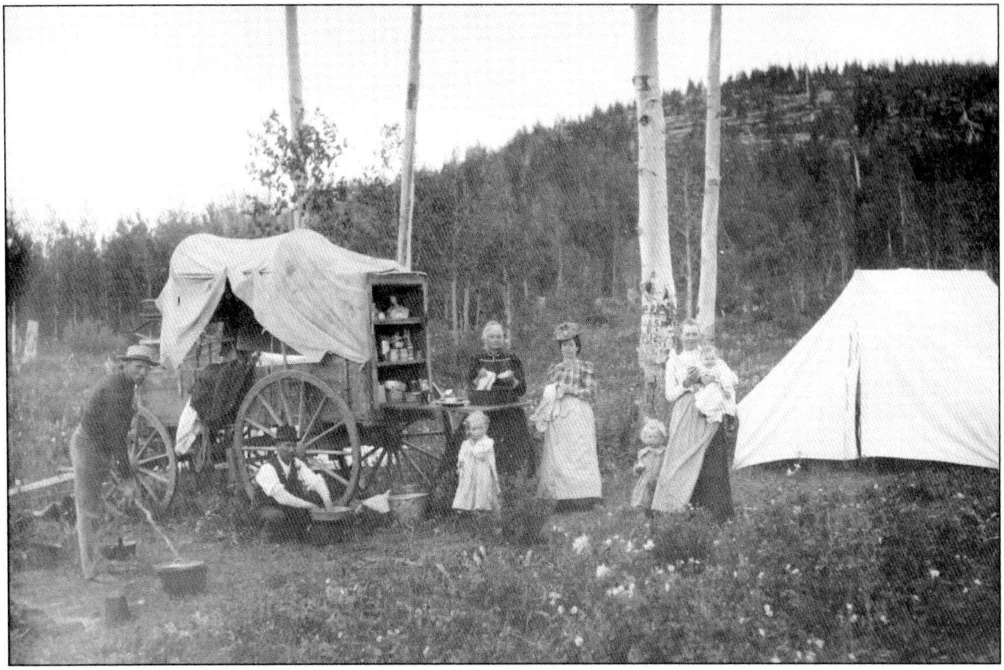

Early Day Camping. The people of Craig often retreated to the mountains for camping. From left to right are Byron Cooper, Archie McLachlan, Amanda Cooper, Ossa Cooper, and Cora McLachlan posing with children at their campsite.

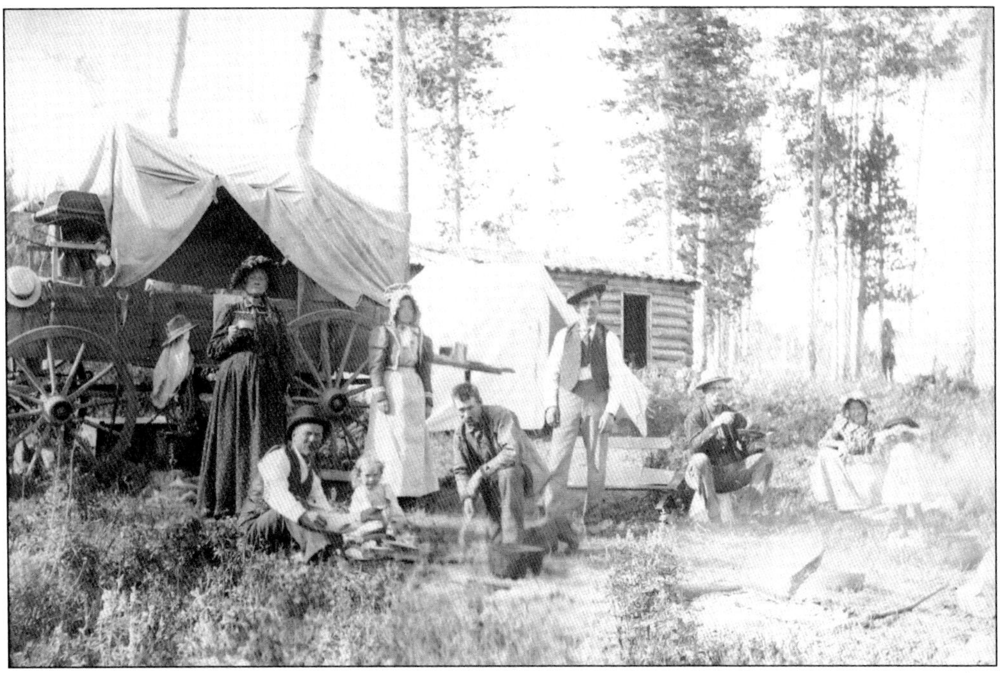

Camping at Hahn's Peak, July 4, 1900. The McLachlan and Ranney families camped at Hahn's Peak for the holiday. From left to right are Amanda Cooper, Archie McLachlan and his little daughter Alma, Marcia (Ranney) Smith, Lowden Ranney, unidentified, Byron Cooper, and Cora McLachlan with her son Hunt.

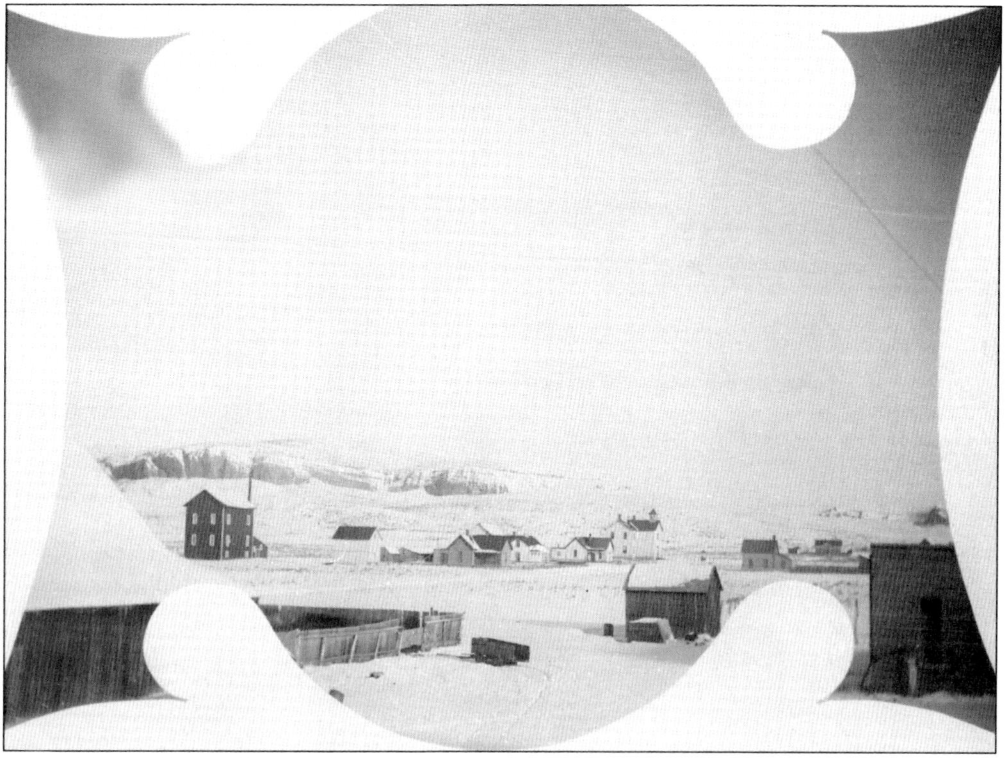

Snowy Day in Early Craig. This image shows the northwest section of Craig around 1898, just north of present-day West Victory Way.

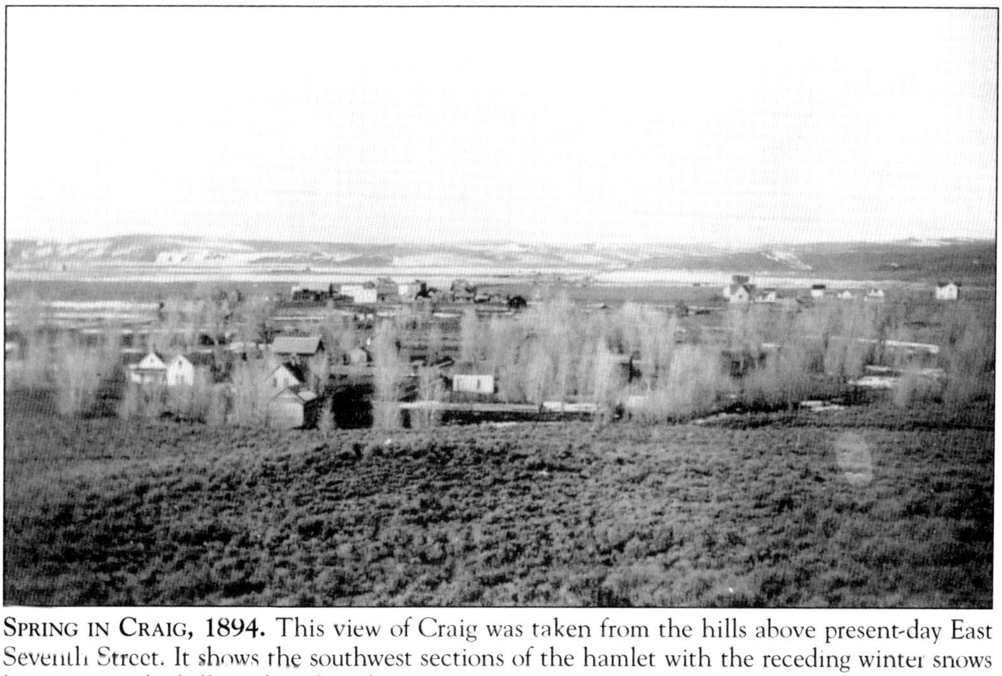

Spring in Craig, 1894. This view of Craig was taken from the hills above present-day East Seventh Street. It shows the southwest sections of the hamlet with the receding winter snows lingering on the hills at the edge of town.

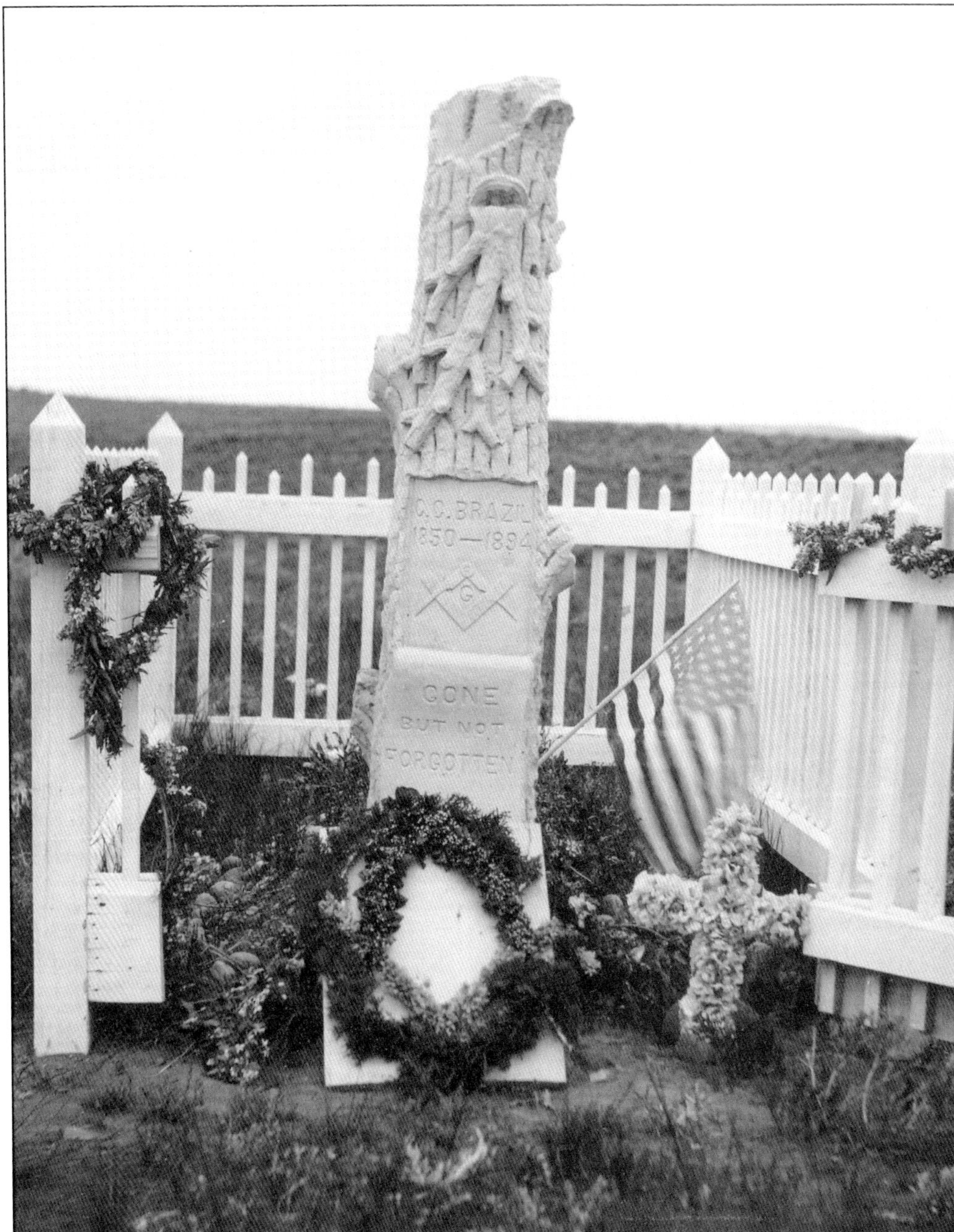

Headstone for a Respected Pioneer. C.C. Brazil was an area pioneer and respected citizen of Craig. When he died in 1894 at the age of 44, this impressive monument was erected over his grave. Weighing 1,100 pounds, it represents the trunk of an oak tree with a center scroll inscribed with "C.C. Brazil 1850–1894. Gone, but not forgotten." This 1896 photograph was taken after the memorial's completion. (Courtesy Tread of Pioneers Museum; photograph by D.W. Diamond.)

OUTLAWS HARRY TRACY AND DAVID LANT, 1898. These photographs were taken in front of the Royal Hotel in Craig after the outlaws' capture in March 1898. The two men had escaped from the Utah State Prison the previous fall and fled to the remote Brown's Park area northwest of Craig. A posse followed them into the Park and Tracy murdered Valentine Hoy, a member of the posse. Sheriff Neiman and Undersheriff E.A. Farnham, who led the posse, were taking the men to the county jail at Hahn's Peak (90 miles away). Lant is pictured at right; Tracy is below. (Both photographs by Amos S. Bennett.)

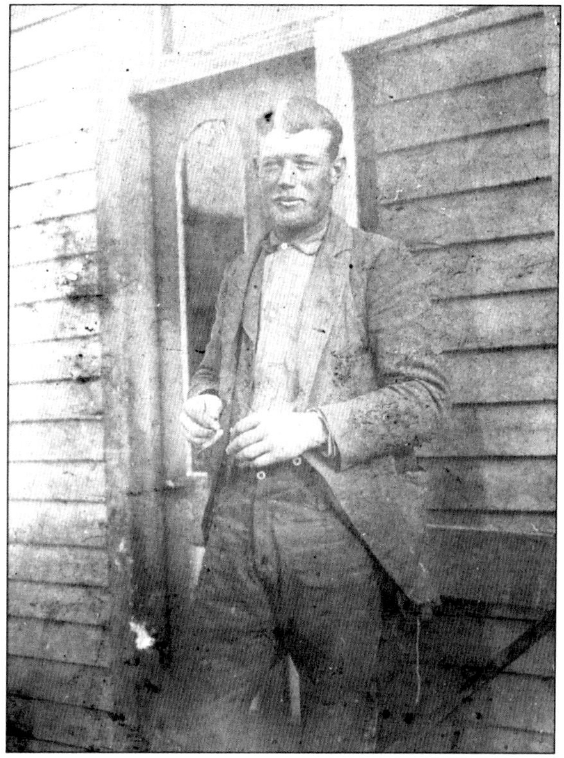

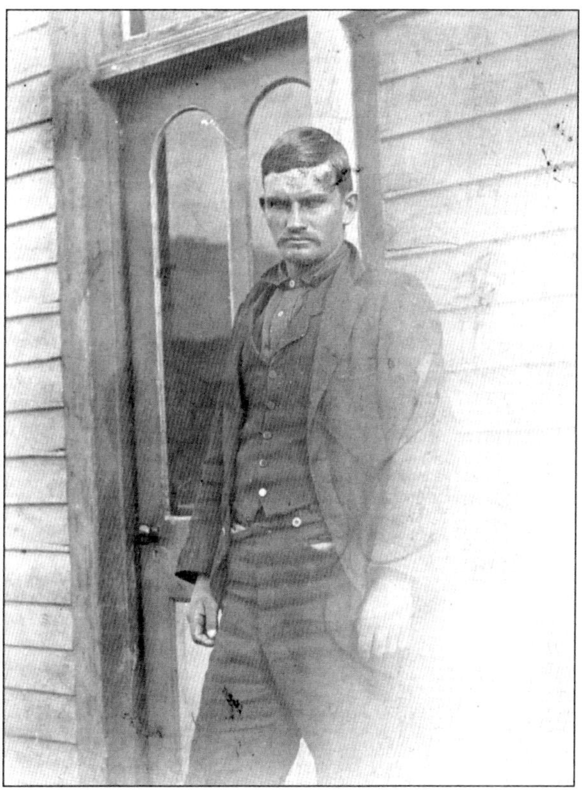

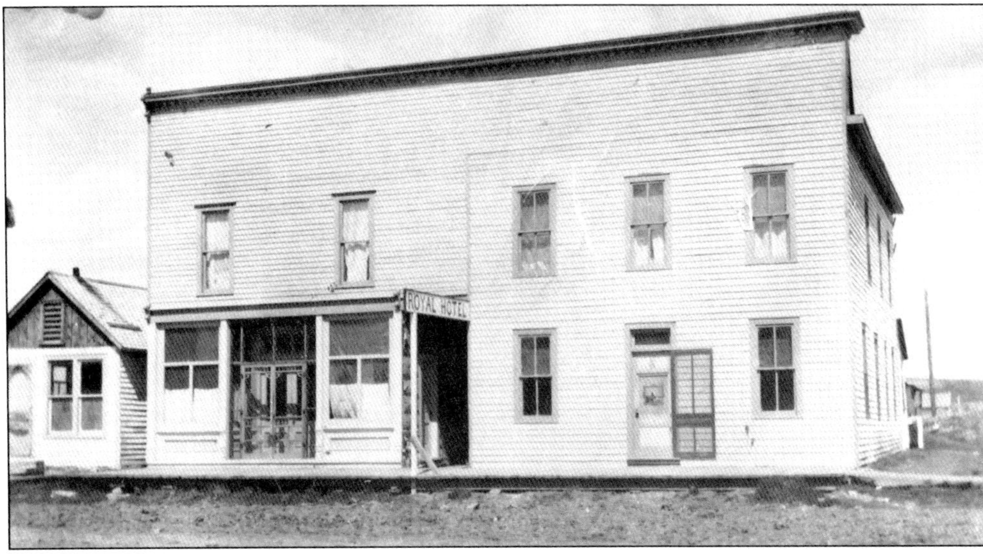

THE ROYAL HOTEL (CRAIG HOTEL), 1910. This was Craig's first hotel, located at the northwest corner of present-day Victory Way and Russell Street. The original structure on the right was built in 1891, and the portion on the left was added in 1909. The hotel was operated for a brief time by Josie Bassett in 1900. Later known as the Alamo Hotel, the building was demolished in 1956.

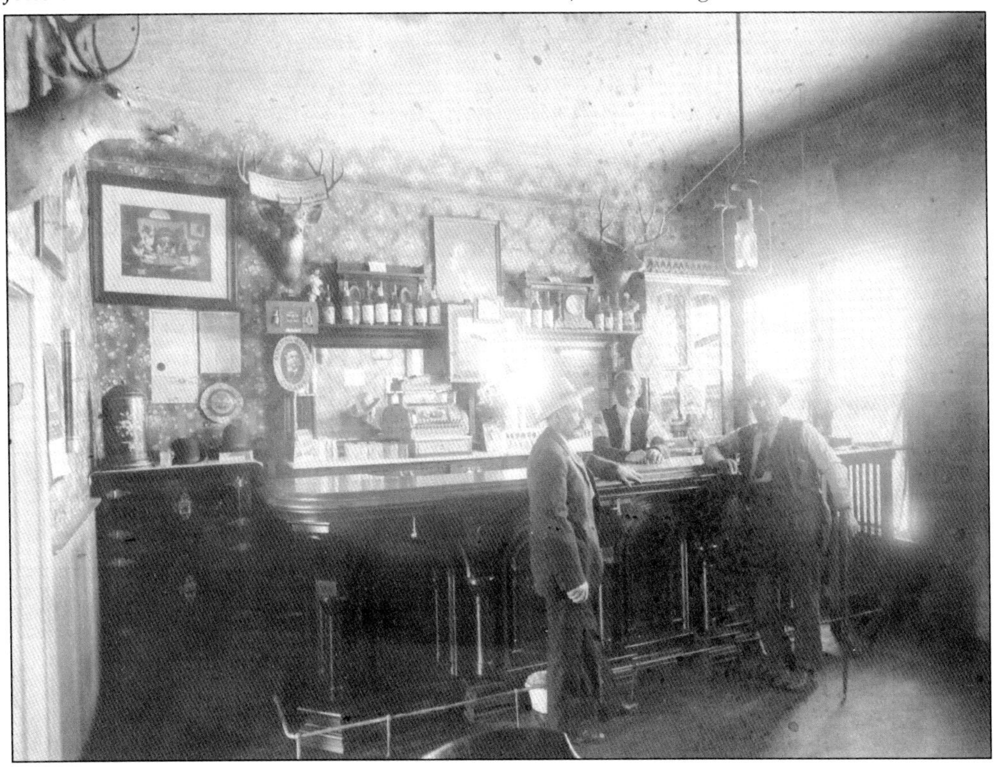

AN EARLY SALOON, C. 1902. Another early business in Craig was the saloon located on Yampa Avenue. This photograph shows the saloon's interior with William H. "Coley" Coleman standing on the far right. When the town was first incorporated, no liquor sales were allowed in the townsite. This restriction was lifted after a few years.

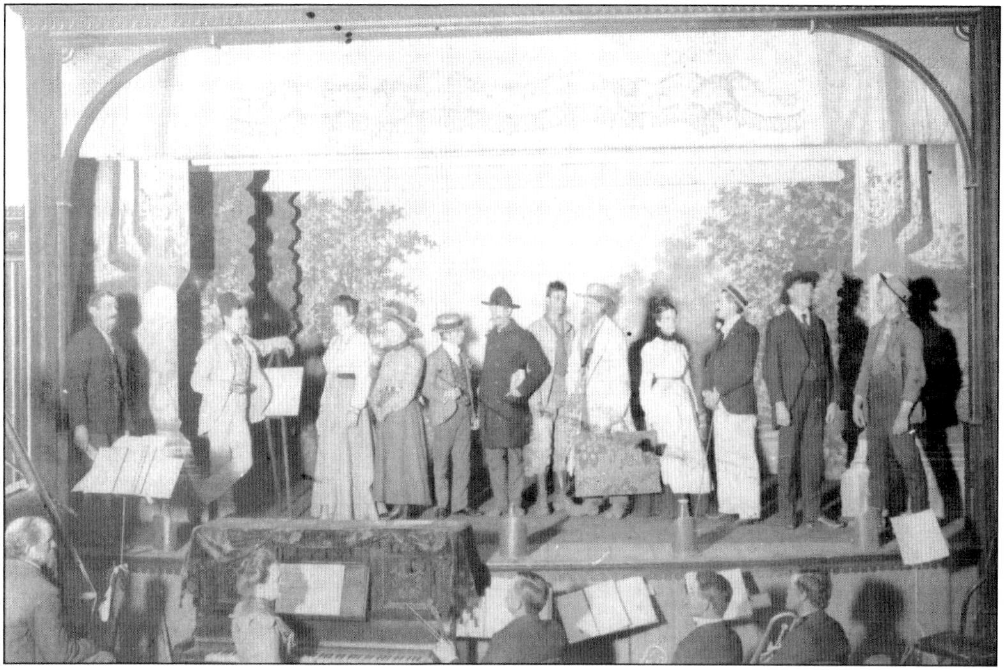

A Craig Dramatic Club Production, 1898. Another sign of a growing community was the construction of an opera house. The cast and musicians of Craig's Dramatic Club posed for their photograph after a performance in the new Opera House at 520 Yampa Avenue. This building was constructed in 1897 after the town hall burned at that same location in February 1896. Among those identified in this photograph are musicians Ernest Kohler, George Woolley, Mary Collum, and George Humphrey. On stage were Alec S. Robinson, Florence Seymour, Frank Ranney, Walter Spencer, Ed Darnell, Blanche Tilton, Clyde Downs, Myrtle Bryan, Fred Downs, and Charlie Ranney.

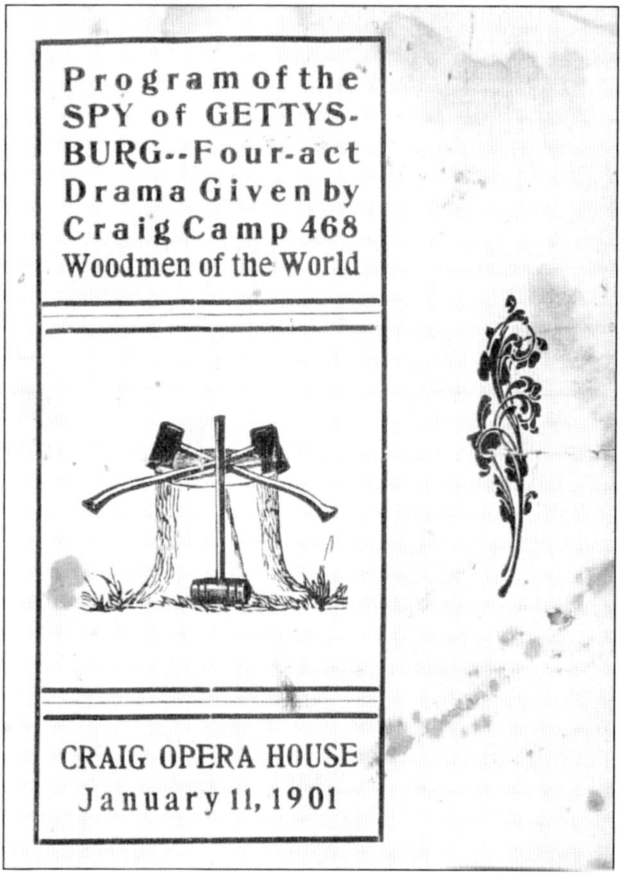

Program of the SPY of GETTYS-BURG--Four-act Drama Given by Craig Camp 468 Woodmen of the World

CRAIG OPERA HOUSE
January 11, 1901

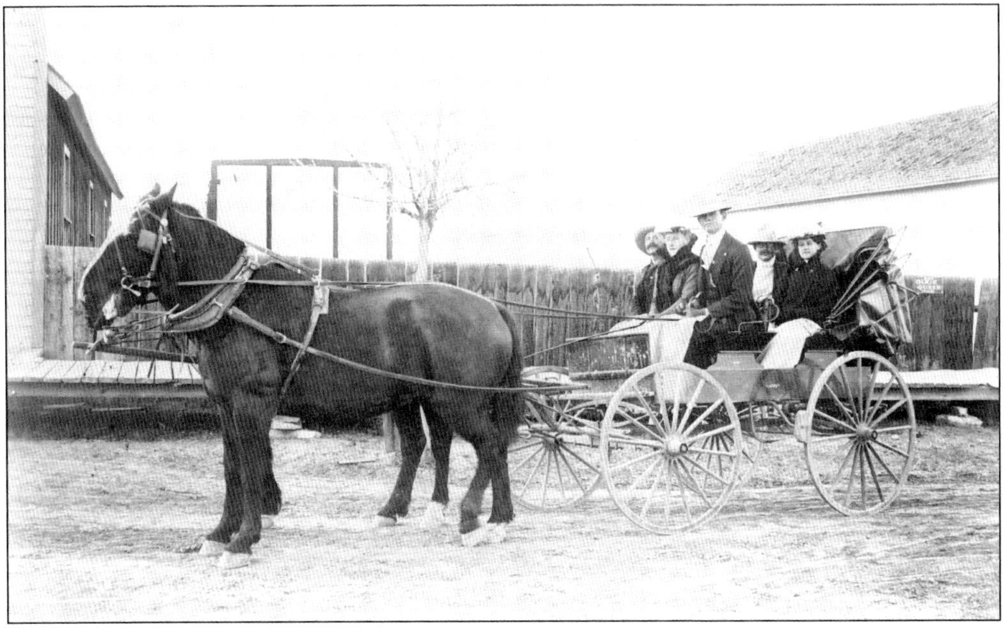

Election Day Outfit, 1900. Bundled against the crisp morning air on election day, a group of Craig residents are pictured on Yampa Avenue just south of the J.W. Hugus store after exercising their civic rights. The only three identified are Tuffy Towers and Ollie Humphrey in the front seat and Ted Hughes, with glasses, in the back. (Photograph by D.W. Diamond.)

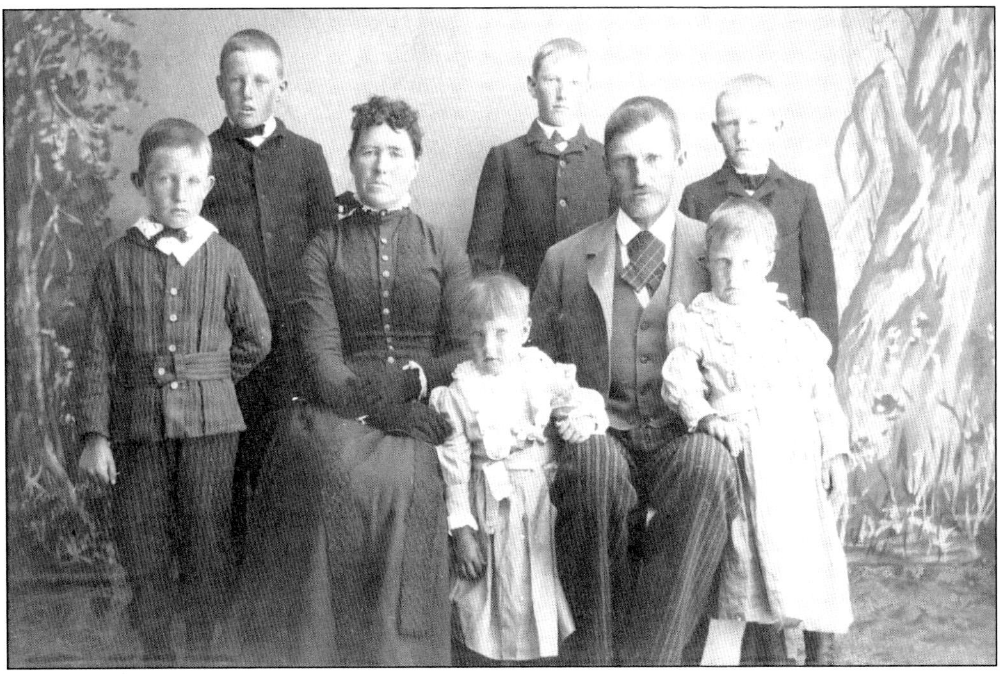

Whispering Green and Family, c. 1900. Robert H. Green (known as "Whispering Green" because of his loud voice) came to the area with his family in 1886, homesteaded east of town, and later bought a ranch west of Craig. Green posed with his wife, Sarah, and their freckle-faced children. (Photograph by A.G. Wallihan.)

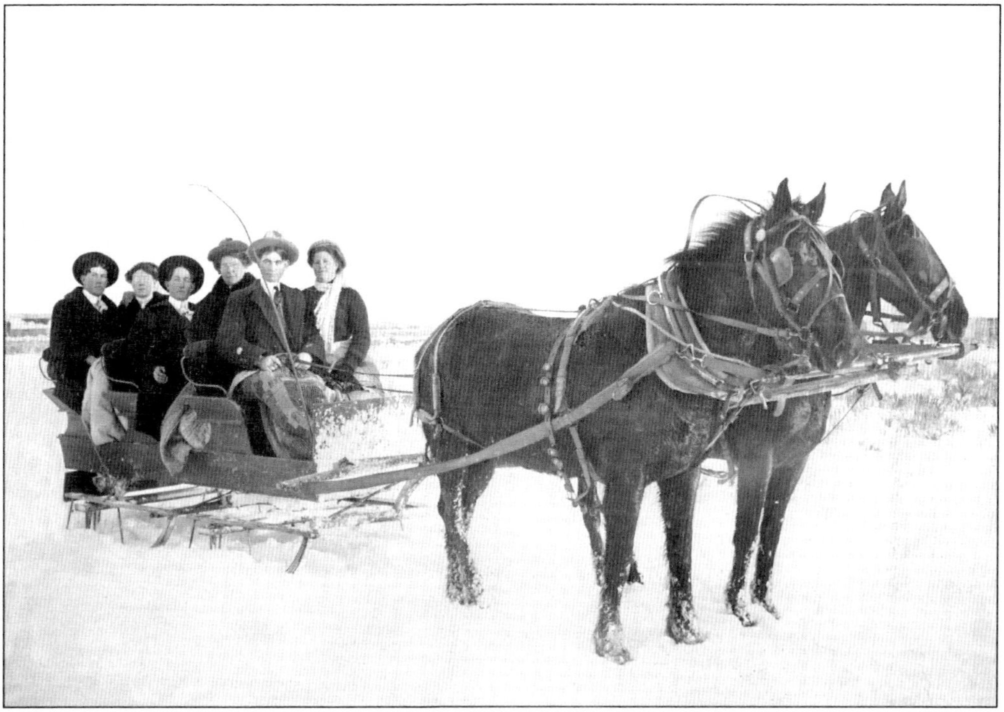

Six in Sleigh, 1904. Pictured here is a sleighing party, including (from left to right) Frank and Ollie Humphrey and Mr. and Mrs. Tuffy Towers. The couple in the middle is unidentified. (Photograph by D.W. Diamond.)

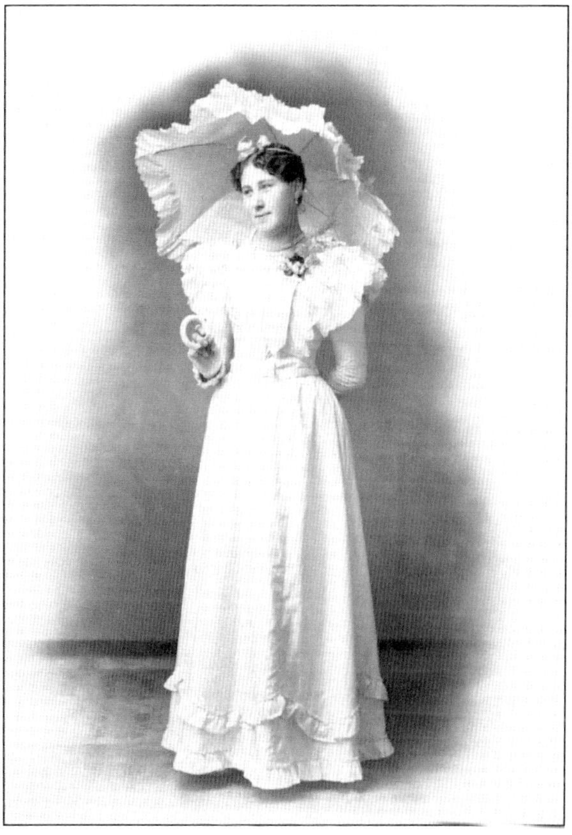

Valley Sweetheart, September 1897. Maud Haughey came to the area with her family in 1886 when they homesteaded just east of town. In addition to school and chores at home, Maud found ample time to socialize; she was a well-liked young person in the community. In 1897, she was elected by popular vote to represent all of Routt County (which included present-day Moffat) at the Festival of Mountain and Plains in Denver.

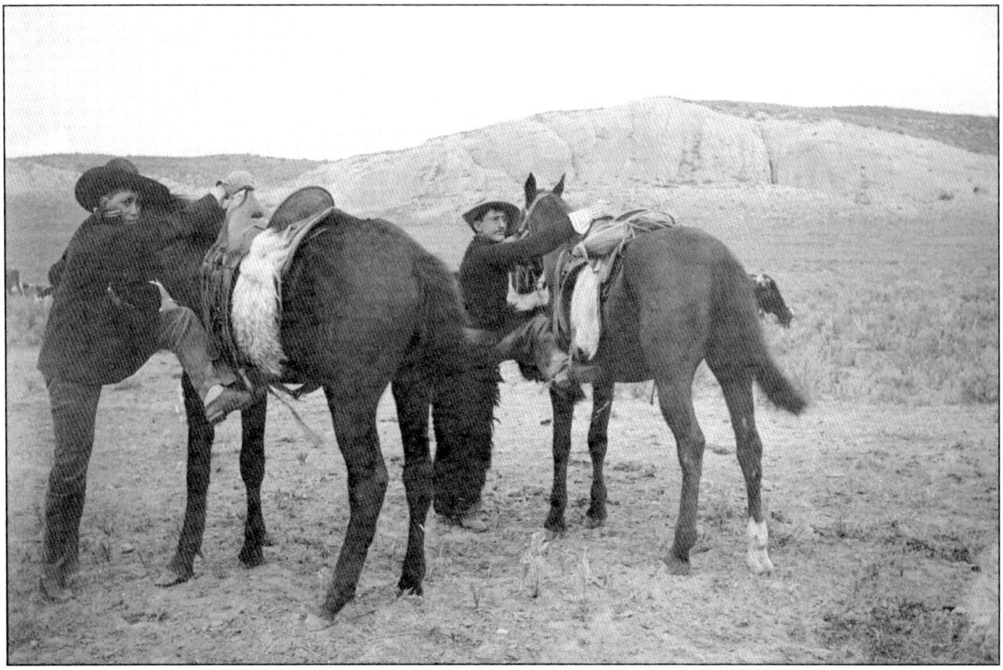

Craig Cowboys, c. 1902. Jim Robinson and Ted Hughes posed for their photograph as they took care of cattle west of Craig. Today, the location is about three blocks north of City Market on Pershing Street. Cattle free-grazed (grazed on open rangeland) here prior to the homesteading era and were a major contributor to the local economy in the first years of the community. Note the Sandrocks in the background. (Photograph by D.W. Diamond.)

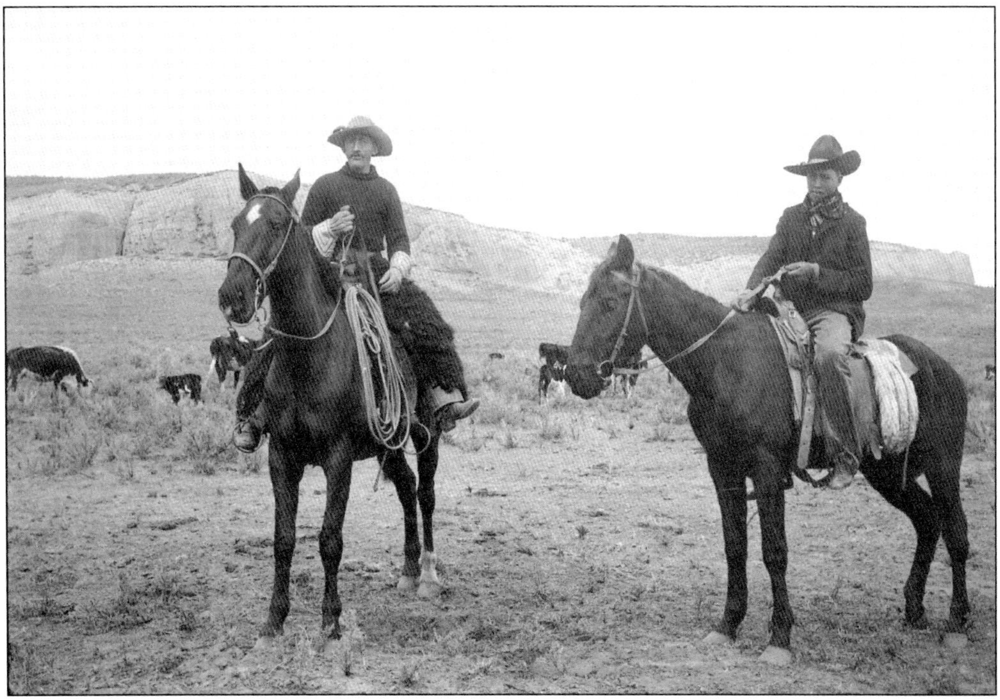

CRAIG, COLORADO
December 22, 1902.

DEAR SIR:

At a meeting of Stockgrowers, held in Craig, Dec. 20, 1902, to discuss methods of exterminating gray wolves, it was decided to call for contributions from the Stockmen of Routt County to form a fund from which to pay bounties. The amount of assessment was fixed at Two Cents per head of live stock in the county, and we trust you will see the urgent need of sending your contribution at once.

Please send to the Bank of Craig, marked—"Wolf Fund."

Yours Respectfully,

E. L. HODGES,
H. H. BARNARD,
R. R. HALL.

The Threat of Wolves, 1902. Routt County stockmen assessed a fee of 2¢ per head of livestock to create a fund to help exterminate wolves.

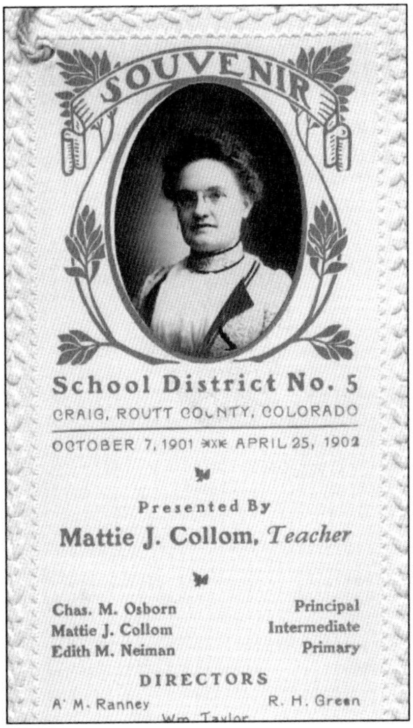

School Souvenir Card, 1901. This elegant souvenir card was presented by Craig teacher Mattie Collom to all 27 of her students listed in the card.

Early Craig School, February 1900. Mattie Collom stands by her desk in her school room in the Craig School, which was built on Breeze Street in 1892. Mattie taught for several years before her marriage to Rev. J.H. Singleton of Golden, Colorado.

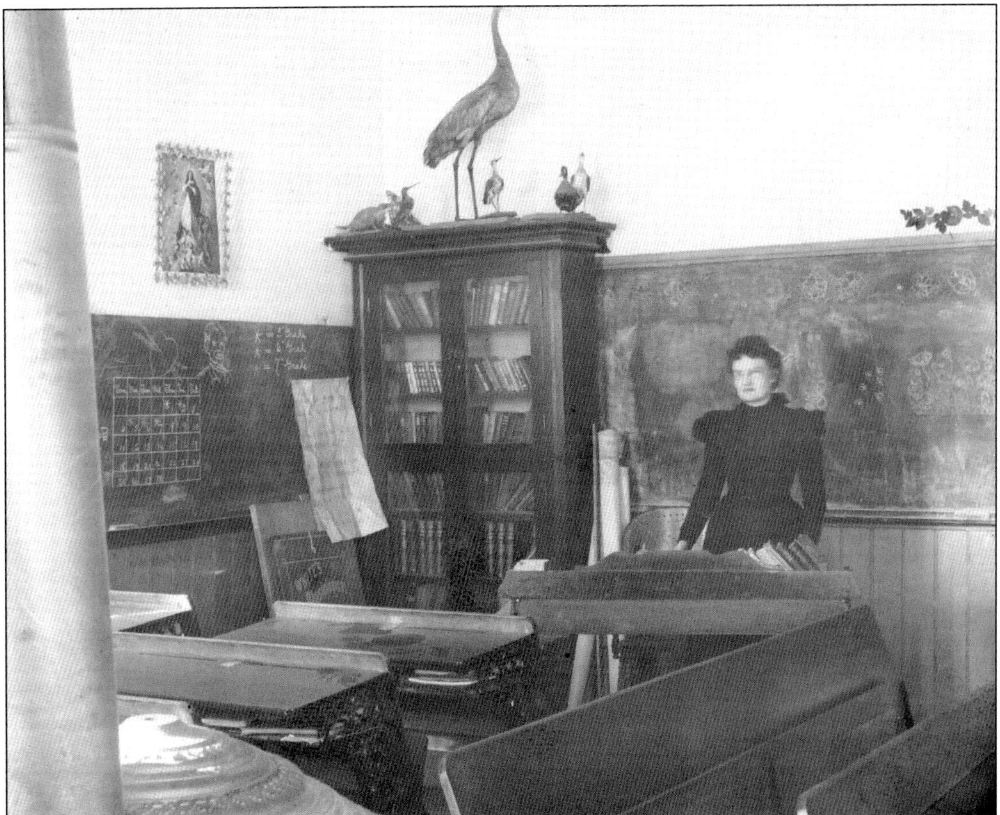

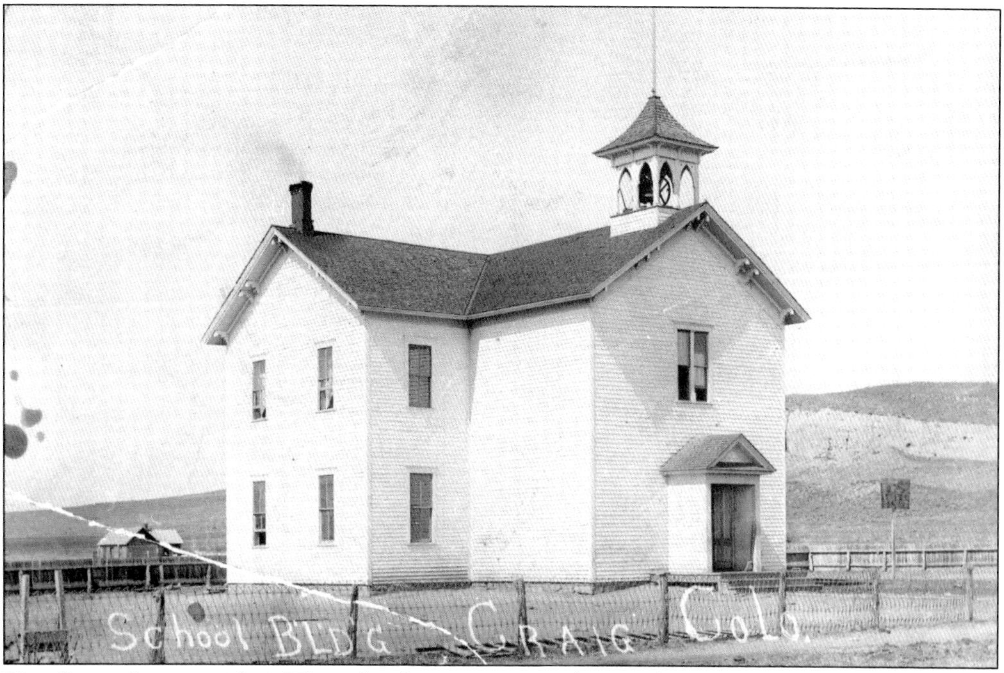

THE CRAIG SCHOOL, 1910. This school was constructed around 1892, and served the community until it was replaced by a large brick school (later known as the Breeze School) in 1916. The photographer, Lewis Hellebust, lived in Craig for a year. (Photograph by Lewis Hellebust.)

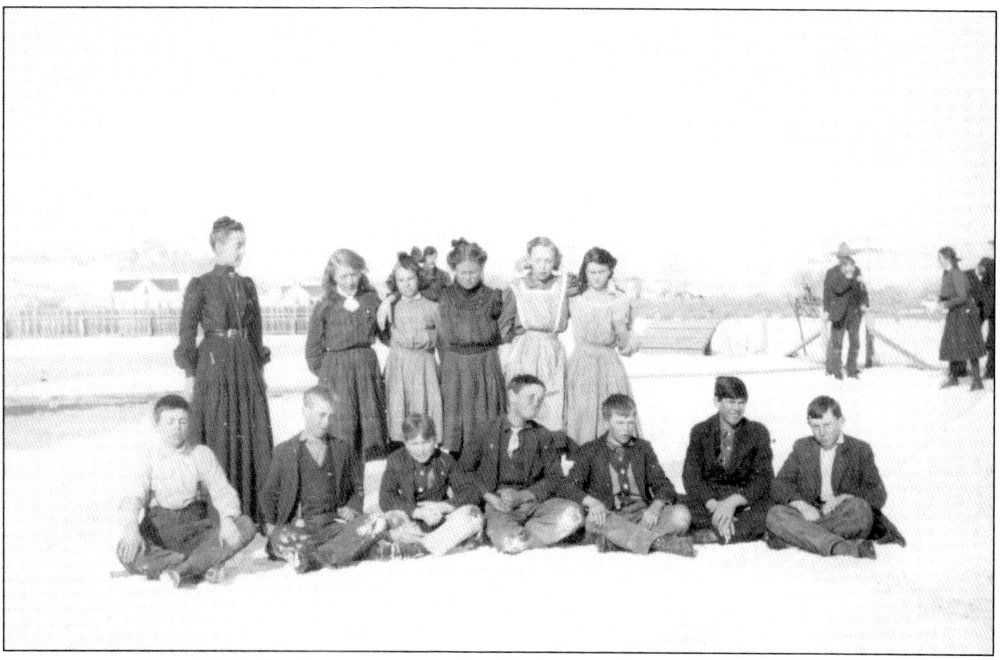

SCHOOLCHILDREN, 1908. Students are pictured with their teacher in the schoolyard at Breeze Street. The only student identified is Pearl McKay (Hamilton) standing at the far right. (Photograph by Robert Mark Richardson IV.)

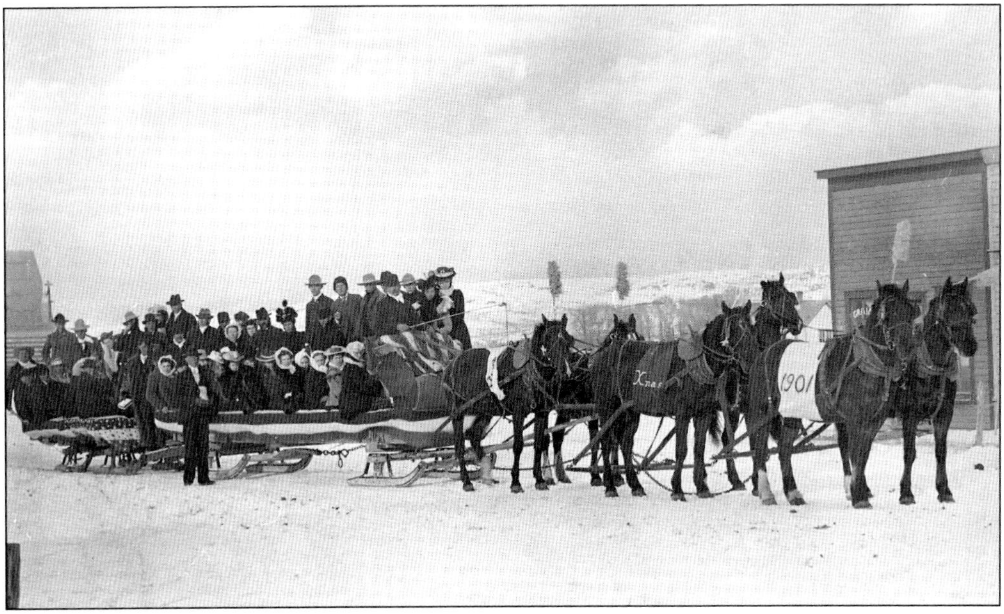

Sleigh Ride, 1901 and 1902. Byron Cooper arrived in Craig in the 1880s with his widowed mother, Amanda. He freighted throughout the Yampa Valley for many years. On Christmas Day, he would take the young people on sleigh rides, and on New Year's Day, he would take the adults. The rides were a wonderful diversion and social event in the cold snowbound months when travel was difficult. Byron died suddenly as a result of the measles in 1906; he was 49.

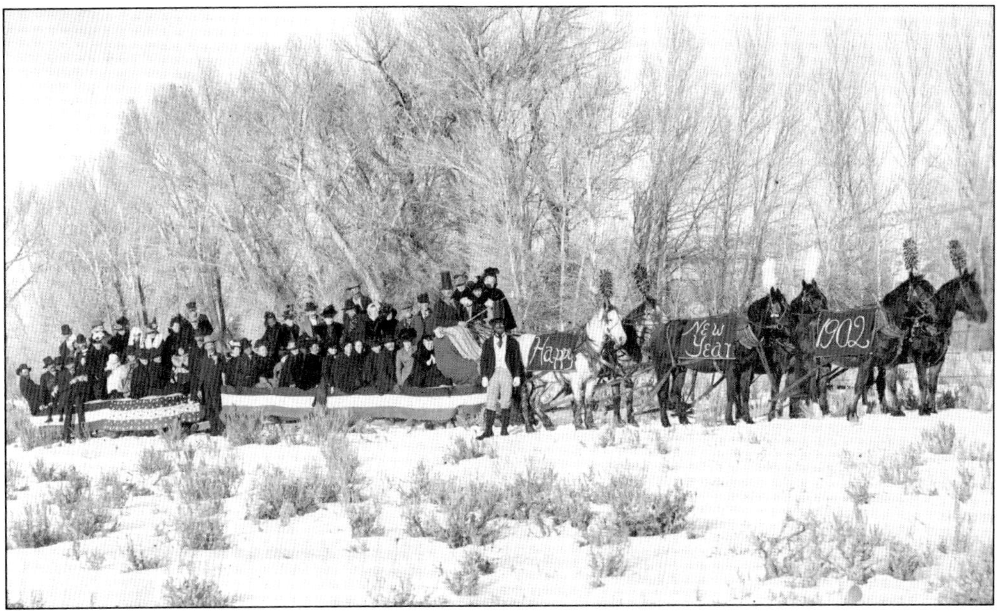

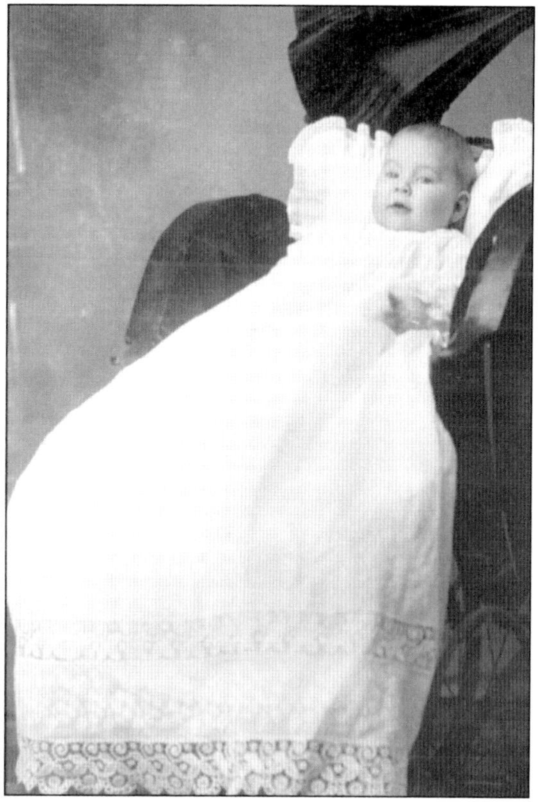

Maude Cooper, 1903. Born to Byron Cooper and Ossa Haughey Cooper, Maude lived her entire life in Craig. At the turn of the century, when she was born, photographer D.W. Diamond was running advertisements highlighting his portrait work of the youngest members of the community.

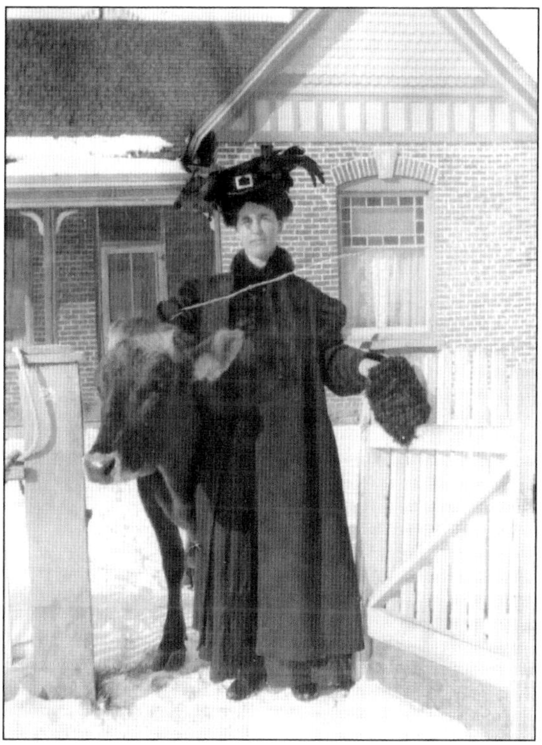

Prized Possession, 1910. Dressed in her finest, Myrtle Coulter posed with her milk cow in her front yard on Yampa Avenue. Her young nephew was photographer Robert M. Richardson IV. (Photograph by Robert M. Richardson IV.)

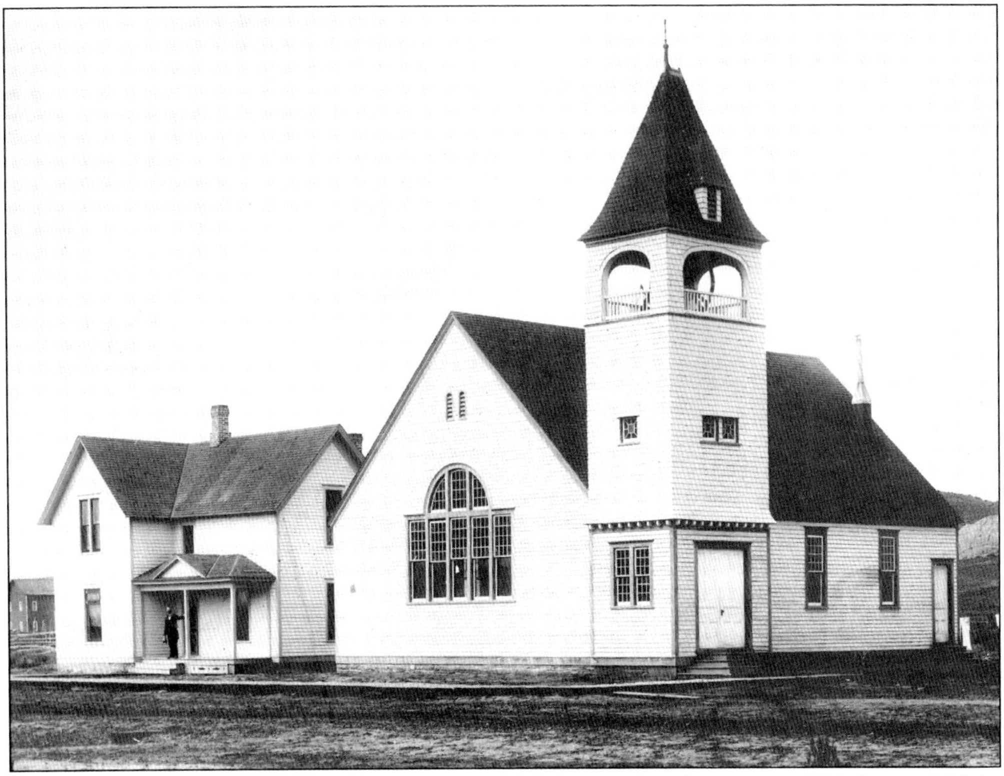

Congregational Church and Parsonage, 1904. Located at the corner of Tucker and Main Streets, the church was completed in 1901, and the parsonage was completed in 1903. The buildings served the church until 1958, when a new church building was erected on Green Street. The original church structure was moved to Green Street and still serves the Methodist and Episcopal congregations. The parsonage was sold in 1958 and moved to the 800 block of Russell Street.

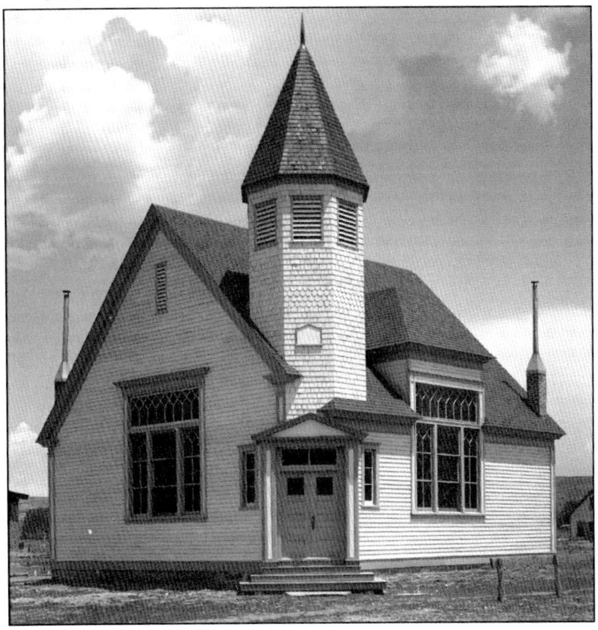

New Christian Church, 1902. The original Christian Church building burned on February 14, 1901, and was rebuilt at the same location in 1902. This structure, with additions, is known as the "Center of Craig" and serves the area as a community center.

THE CITY DRUG COMPANY, 1902. This photograph shows the newly constructed City Drug Company building at 520 Yampa Avenue. Fred Downs, proprietor, stands on the left.

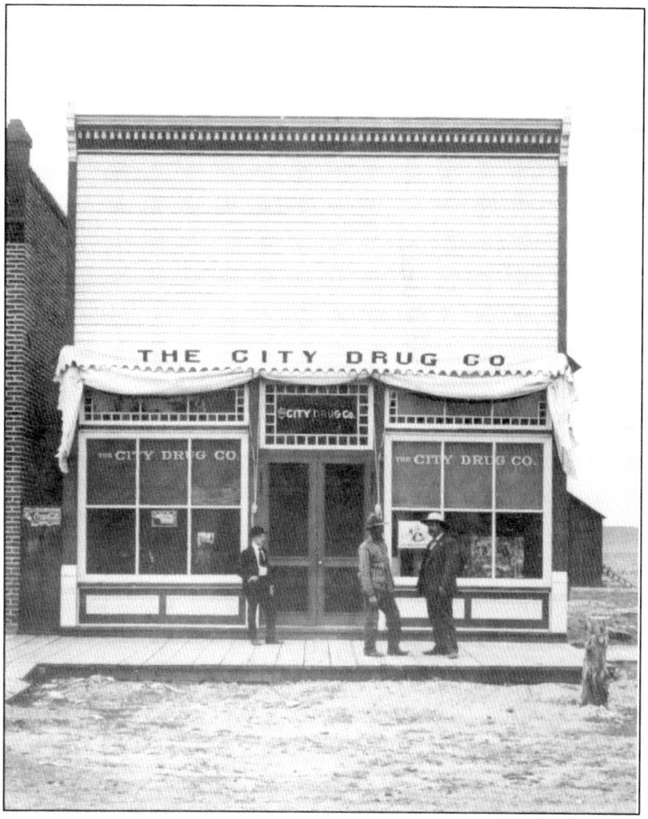

SODA FOUNTAIN, 1902. This soda fountain was an elegant addition to Craig's downtown. The town was entering a phase when businesses of a more frivolous (non-essential items) nature could succeed. Owned by Fred Downs, the fountain boasted eight varieties of soda flavors; it was located inside the City Drug Company on Yampa Avenue.

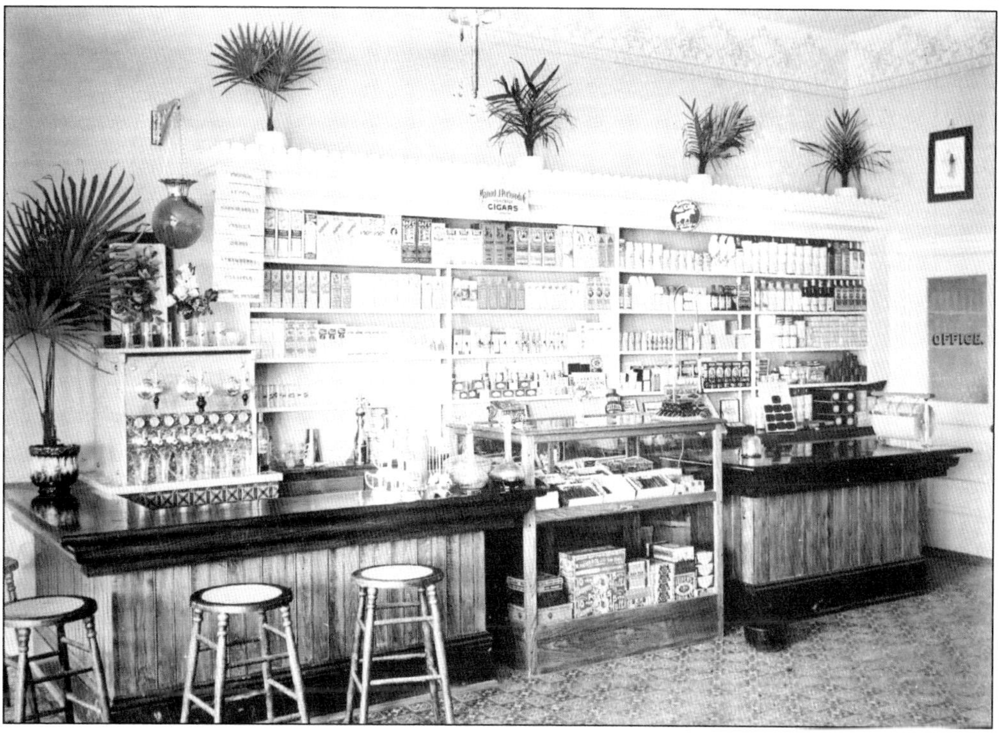

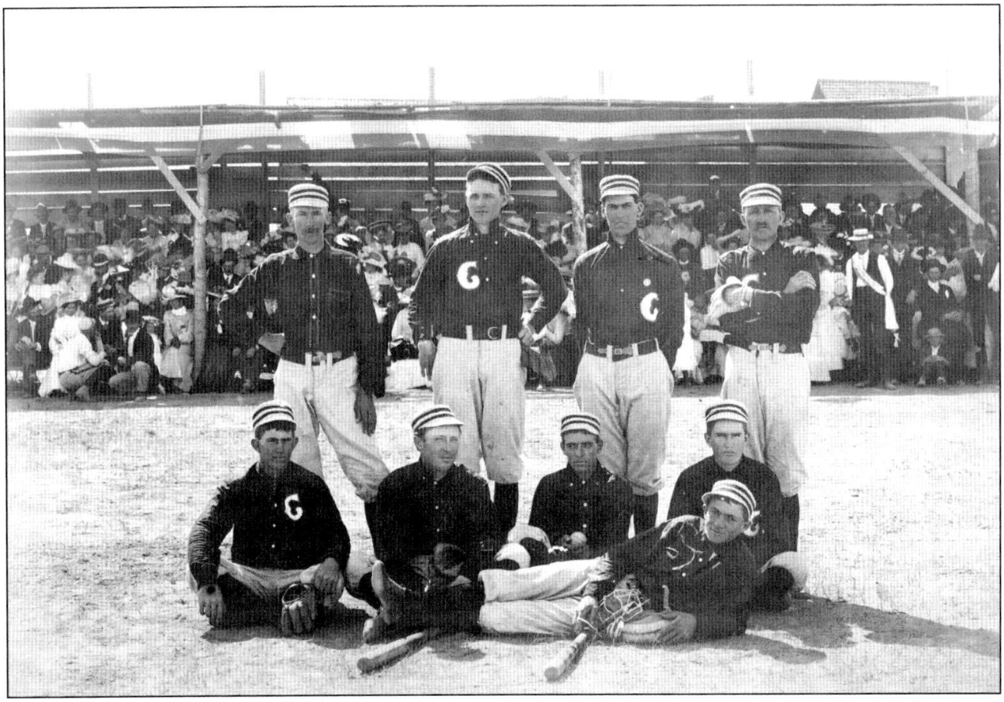

CRAIG BASEBALL TEAM, JULY 4, 1904. Baseball was soon an integral part of the small town's social activities. Teams around the region traveled to compete with one another. From left to right are (first row, seated) Jeff Biggs, Dr. Guy Leach, George Welch, and Carl Van Dorn, with Chub White reclining on his side; (second row) Charles Ranney, Ed Howard, C.S. Merrill, and P.H. Howard.

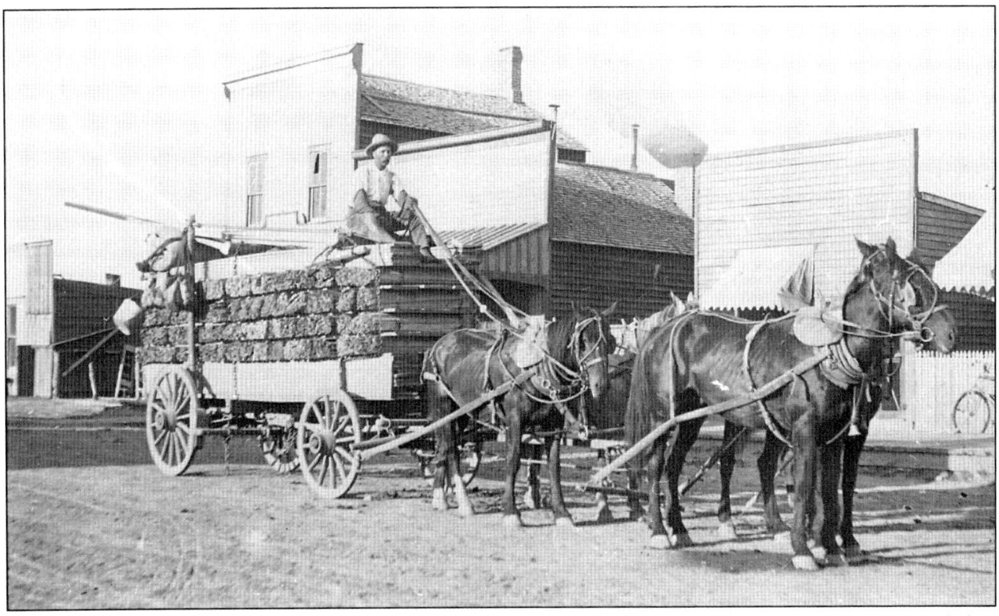

SAWMILL PRODUCTS ON THE MOVE, 1902. Bryon Cooper stopped on Yampa Avenue with his wagonload of lath as he was hauling it from McLachlan's sawmill on Black Mountain to the Cary Ranch near Hayden.

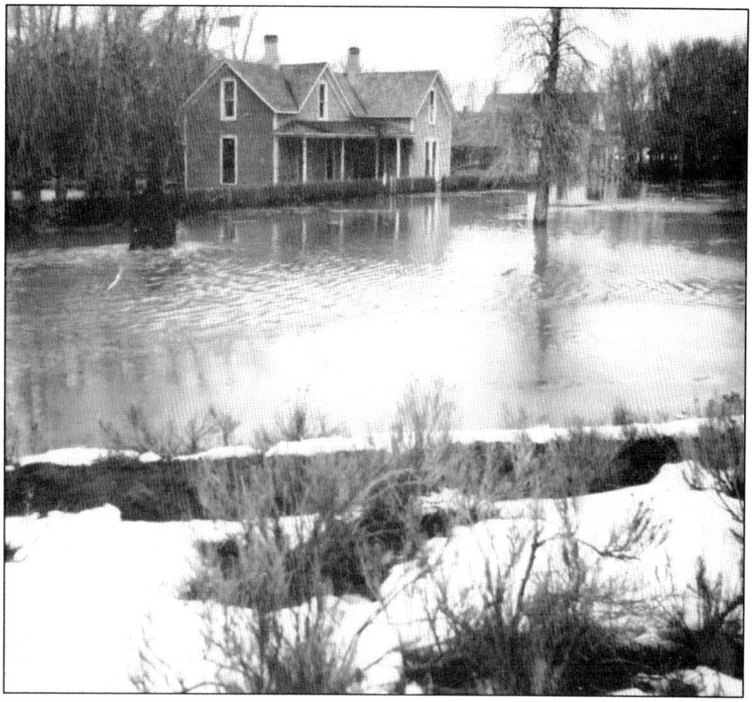

Inundated Homes, 1904. The night of February 23, 1904, brought a surprise to many residents of the Rosedale subdivision in east Craig. An unusual thaw caused an ice jam in Fortification Creek; the water in some places stood 18 inches deep. Men blasted the ice with dynamite, and by the following day, the water had receded. In the 1940s, the creek was finally dredged to avoid more flooding.

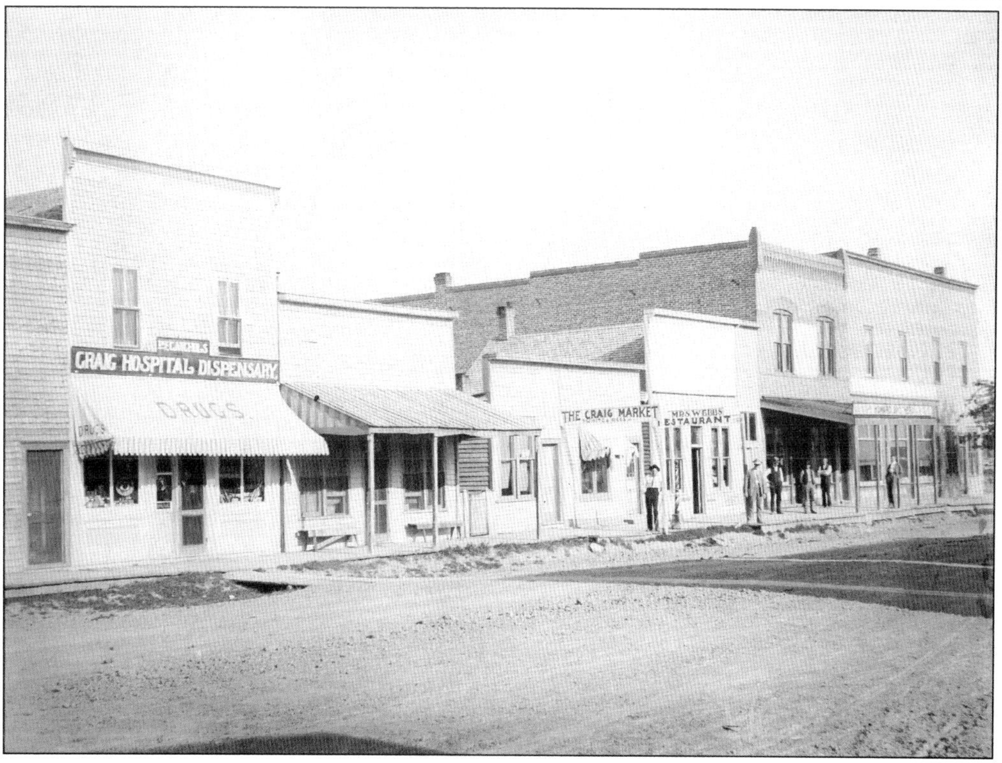

Busy Downtown, 1907. The east side of Yampa Avenue was filled with businesses by this time. A drugstore, meat market, barbershop, restaurant and general mercantile were among the businesses that offered their services to local residents.

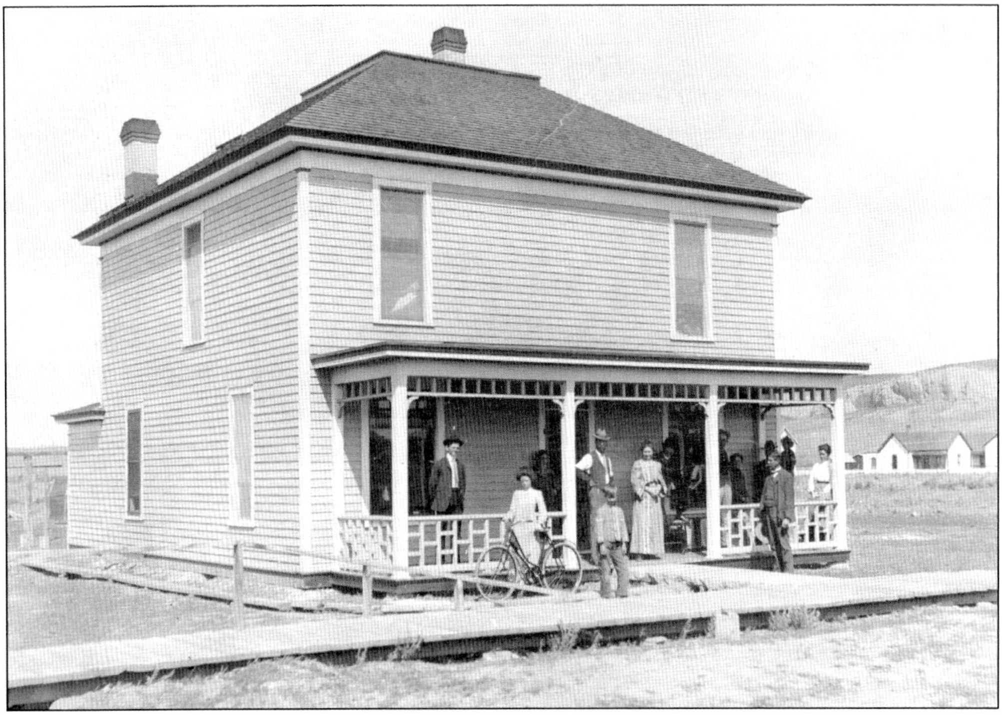

The Baker House Hotel, c. 1903. Cora and Charlie Baker, center, pose on the porch of the Baker House Hotel with their guests. Maude, their daughter, stands with her bicycle in front. The Bakers built this hotel at 538 Yampa Avenue in 1902. It was in use as a restaurant until recently.

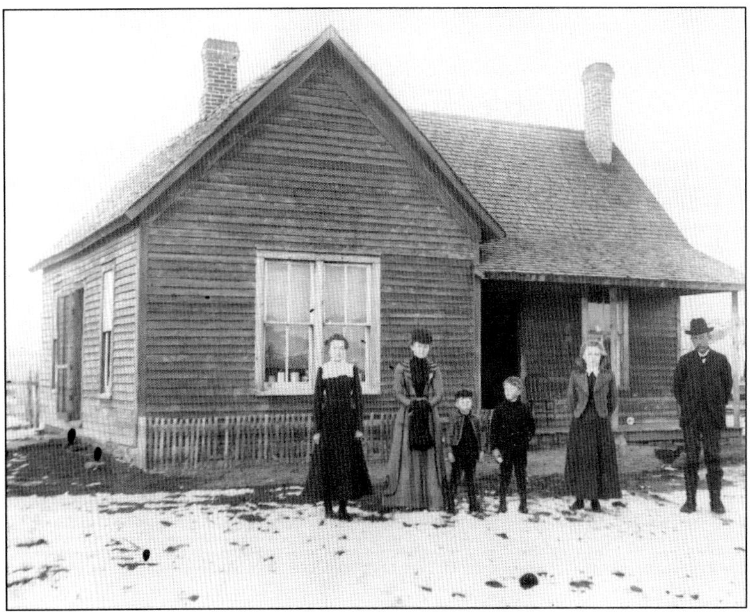

Bryan Family and Friends, 1901. R.V. Bryan poses with his family and friends in front of the Bryan home at the corner of Sixth and Russell Streets. From left to right are Maude Bryan, Josie Bassett with her two sons, Crawford and Herbert (Chic) McKnight, Myrtle Bryan, and R.V. Bryan.

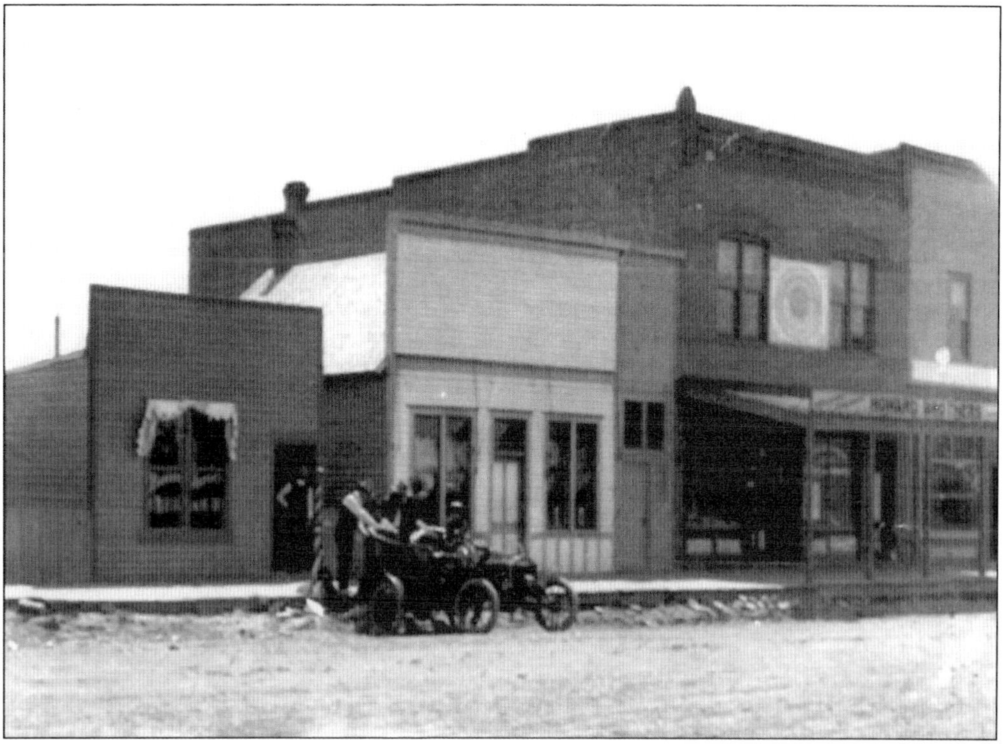

First Automobile in Craig, c. 1904. Parked on Yampa Avenue, this automobile is thought to be the first in Craig. To drive a car to this remote town was quite a feat considering the rough wagon roads of the time. It was a sign of the changes coming in transportation and the opening of the isolated community to larger towns on the other side of the Rocky Mountains.

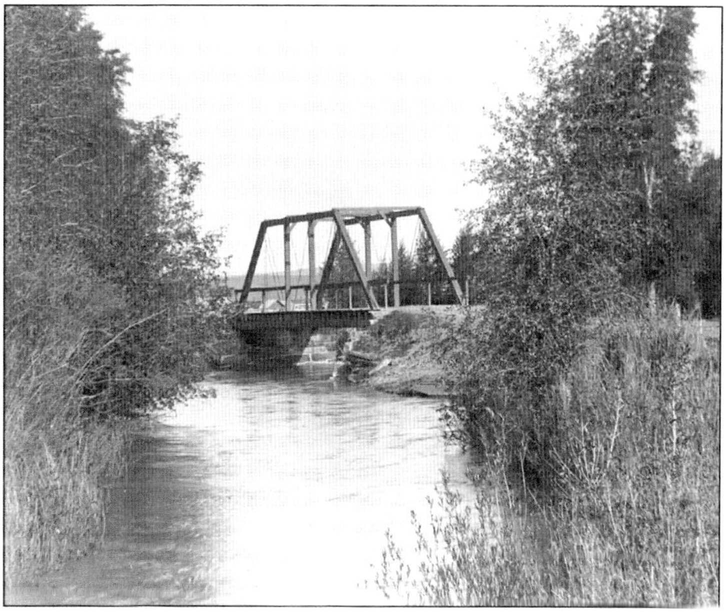

Fortification Bridge, 1900. This bridge was built over the creek near present-day East Victory Way. Travel on the east-west road along the Bear (Yampa) River increased as the region's populations grew. This modern bridge was an improvement for those traveling east along the road to Hayden and to the county seat at Hahn's Peak.

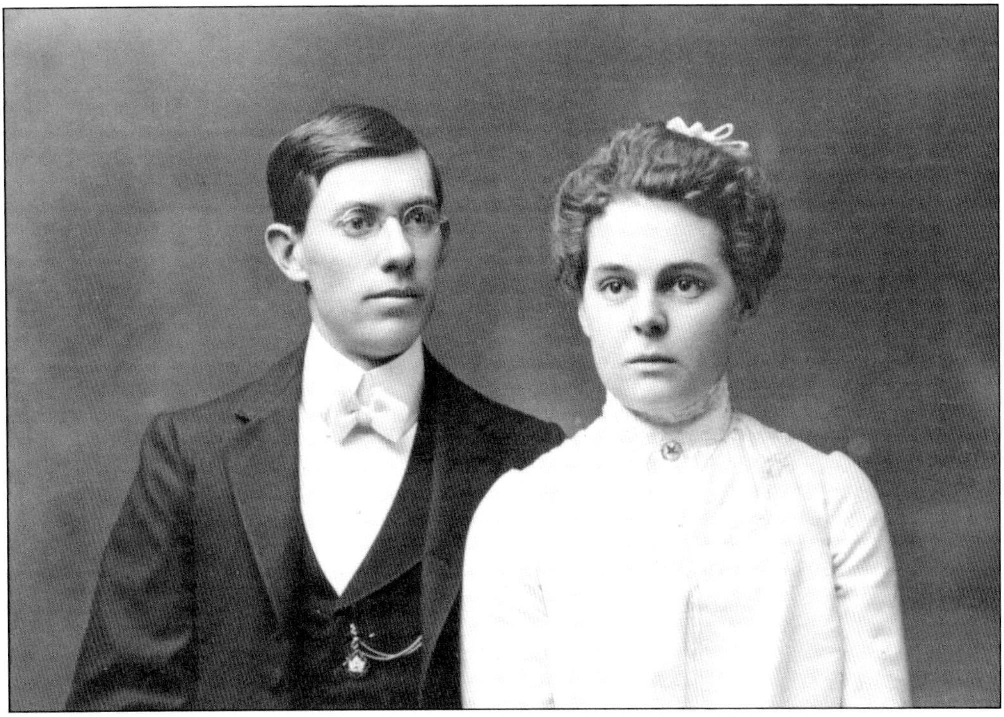

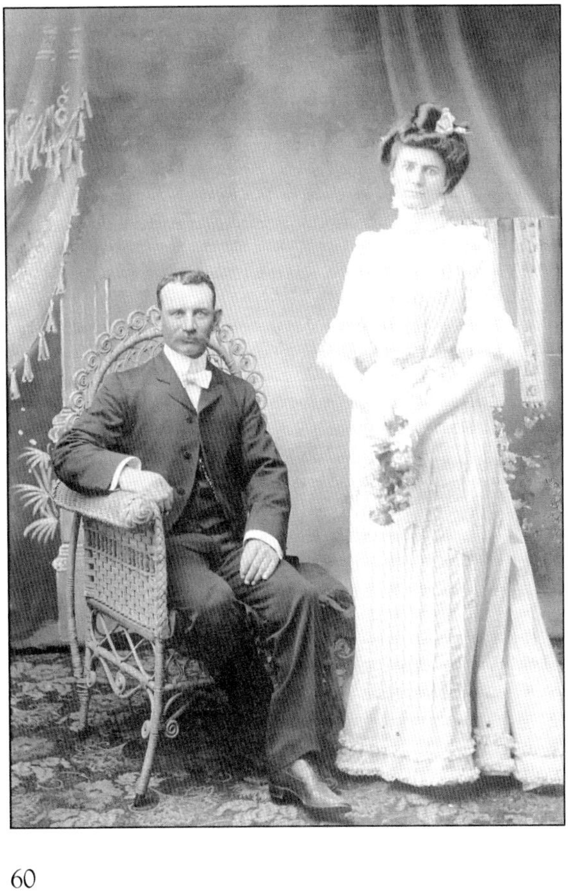

MARY AND CLYDE DOWNS. Mary Taylor came to Craig in 1894 with her parents, Margaret and William Taylor. Clyde moved to Craig with his parents in 1895 and was in the mercantile business in the young community. Mary and Clyde married in 1901 and were active in Craig throughout their lives. Clyde, besides being involved in local commerce, served terms as county commissioner, county treasurer, and Craig city mayor. (Photograph by D.W. Diamond.)

WEDDING, 1902. Cullie Melugin and Ora Biggs were married in 1902 and settled on Waddle Creek. Hard workers in the cattle, sheep, and dairy industry, they later moved to a ranch near town. The spacious house they built is now the parsonage for the Craig Christian Church on West Victory Way. Ora died in 1927, and Cullie later married Ed Walsh. Cullie died in 1988, just before her 105th birthday. (Photograph by D.W. Diamond.)

ARCHIE McLACHLAN FAMILY, c. 1904. Archie and Cora McLachlan pose with their children Alma, Audrey, Edwin, and Hunt. Archie came to Craig from Nova Scotia and homesteaded north of Craig. He also opened a large sawmill on Black Mountain, providing lumber for the growing region. He married Cora Ranney of Craig in 1895. (Photograph by D.W. Diamond.)

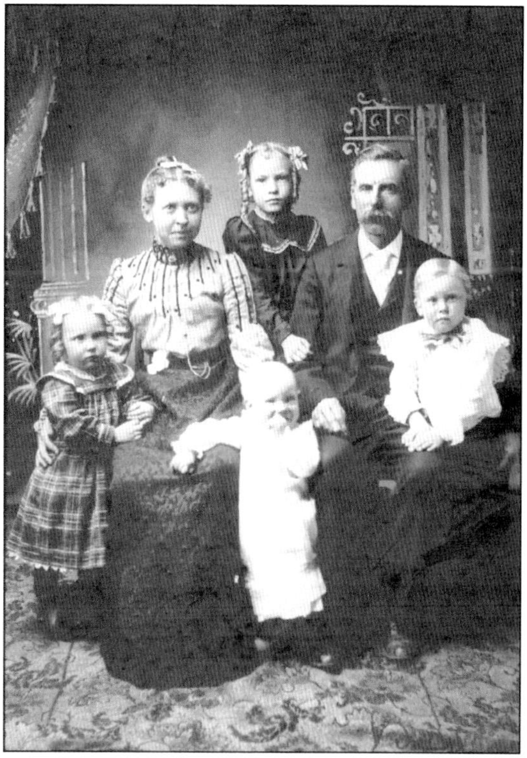

OLD TIMERS, 1910. A large group of men, instrumental in creating this community out of a frontier, gathered at Archie McLachlan's Ranch southwest of Craig. Some of the men present were Robert Kimbley, R.M. Richardson, Oliver Haughey, Lemuel Breeze, Joe Carroll, John Jackson, William Taylor, John Ledford, Alex Robinson, William Rose, Louden Ranney, Frank and Charlie Ranney, Charlie Daniels, R.H. Green, William Tucker, Tom Forkner, and Matt Johnson.

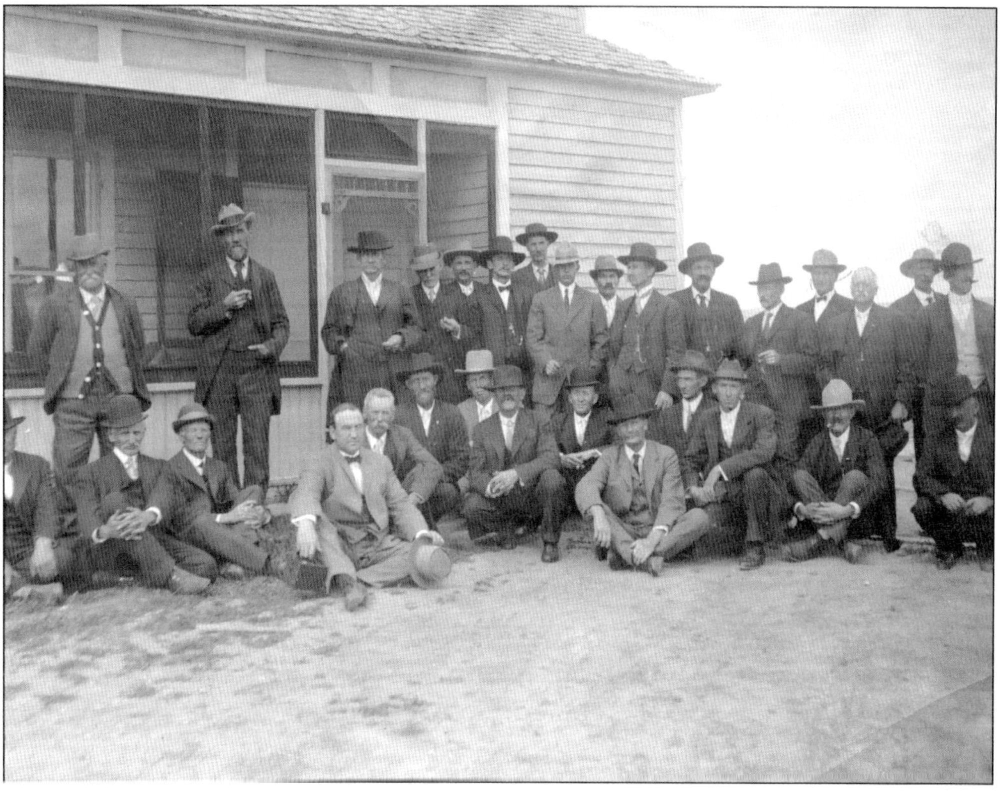

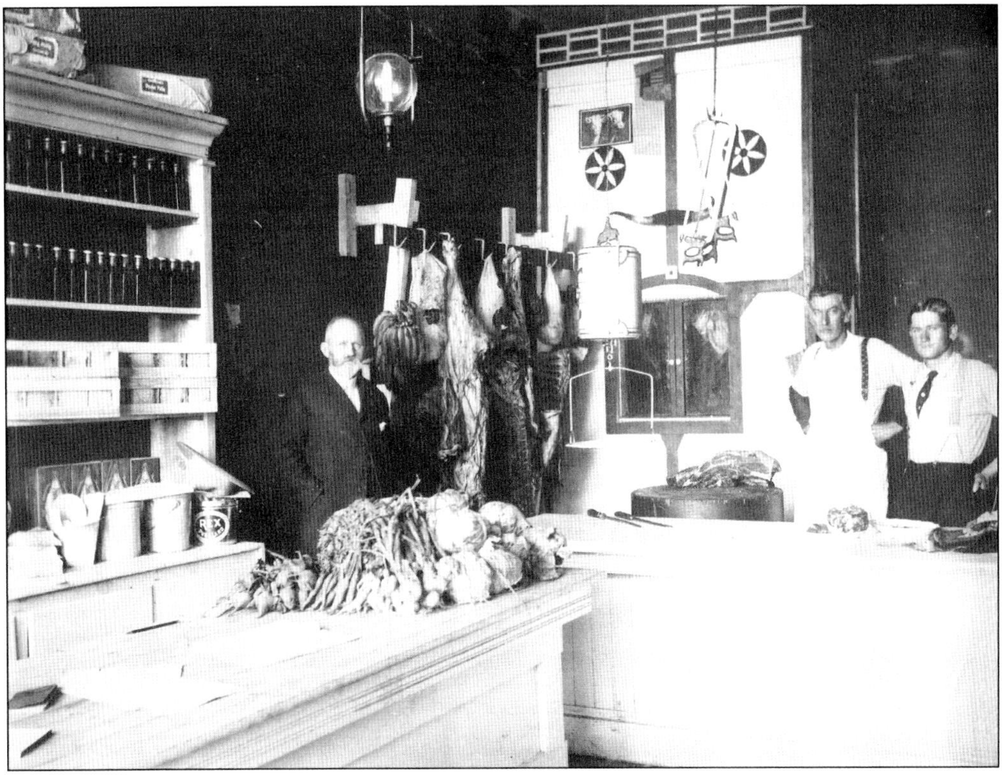

ENTREPRENEURS, C. 1908. The townsmen were just as instrumental as the homesteaders in the development of this rural community. The merchants gambled heavily as they invested in buildings and commodities for the early families of the area. White's Meat Market operated in the 500 block of Yampa Avenue. From left to right are Charles M. White, Wash Held, and Ralph L. White.

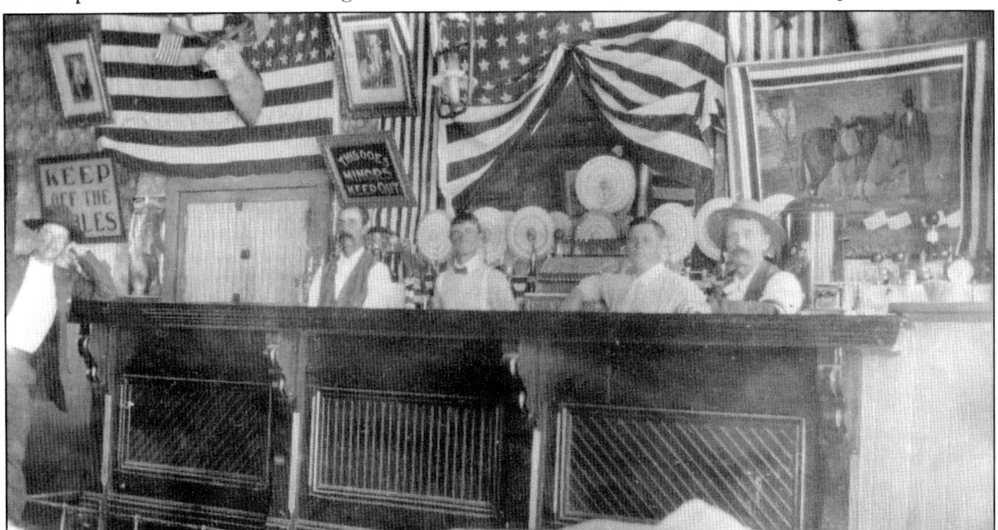

LEDFORD AND KITTELL SALOON, 1910. Saloons were staple businesses for frontier towns. The interior of the Ledford and Kittell Saloon is complete with patriotic flags and warnings for patrons to "Keep Off the Tables" and for minors "This Goes—Minors Keep Out."

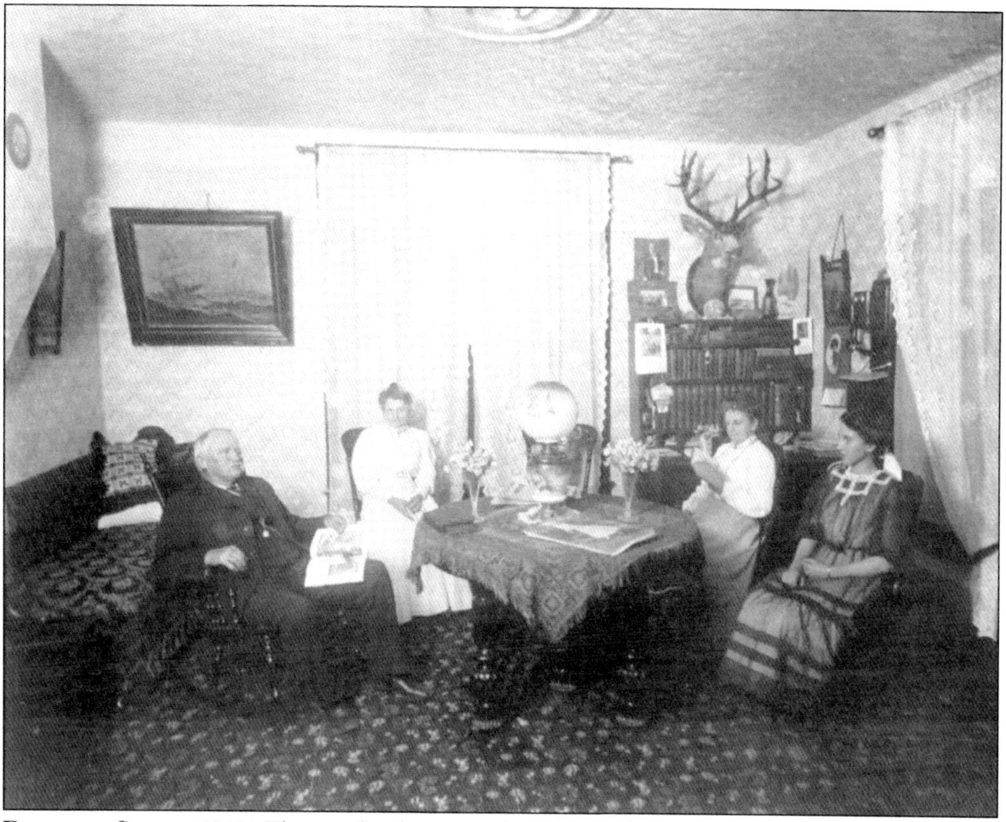

DOMESTIC SCENE, 1907. The rough edges gone from his early pioneer years in a log cabin, William Rose posed in the comfortable living room of his Craig home. His wife, Julia, is seated in front of the window along with her sister Emily Jamison and young Pearl Aiken (White) who was working for the Roses at that time.

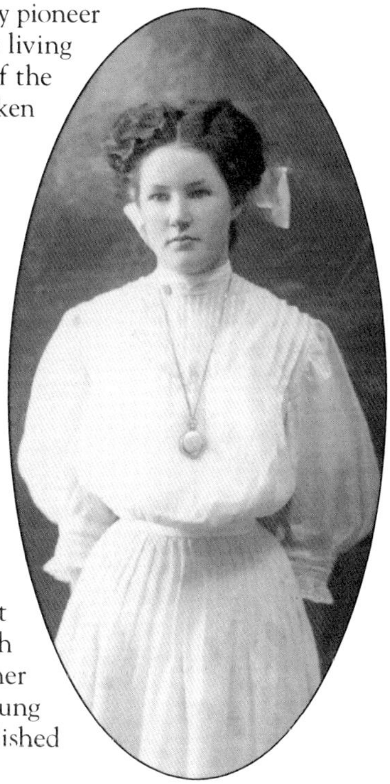

BELLE HUMPHREY MCKEE. Belle, born in 1890, was the first baby girl born to area settlers. Her family homesteaded north of Craig in the 1880s. After the father's death, Belle's mother operated a millinery shop on Yampa Avenue to support her young family. Belle married Charles McKee in 1912, and they established their home in Denver.

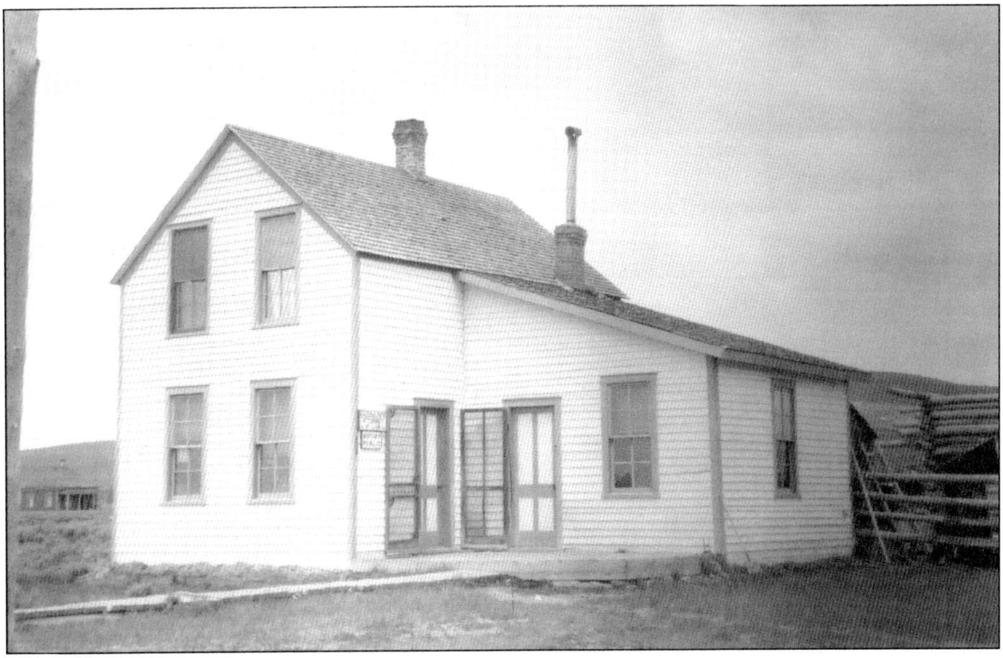

JUDGE JOHN A. COULTER'S RESIDENCE. A Civil War veteran, Coulter moved with his family to Craig from Georgetown, where he had been an active criminal lawyer. He served as a judge and lawyer in Craig for a number of years before moving to Boulder in 1917. His home was located at the southeast corner of West Victory Way and School Street, where the courthouse now stands.

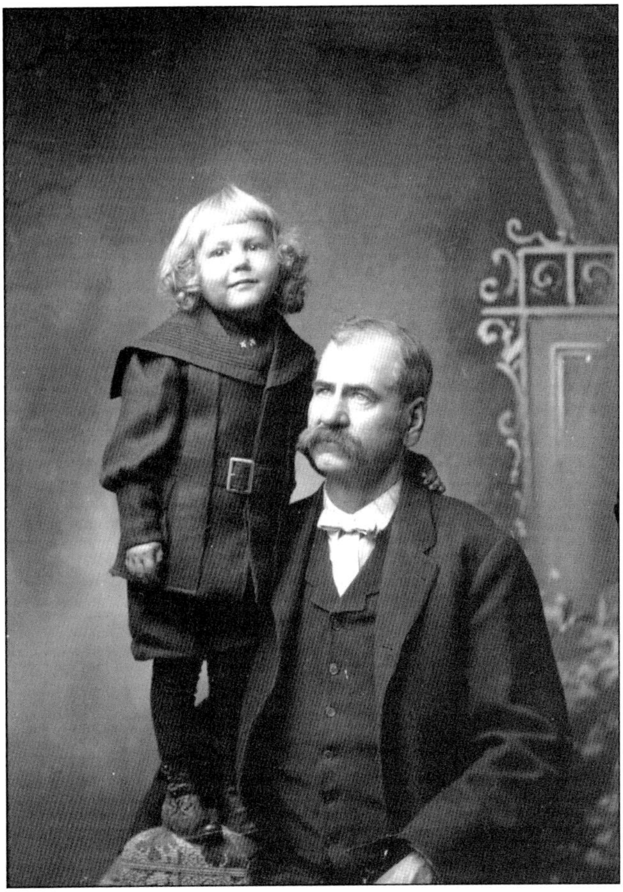

JOHN A. COULTER AND HARMON COULTER RICHARDSON, 1903. Proud grandfather John Coulter posed with his grandson, the child of his stepdaughter Laura Richardson. Craig, like many frontier towns, was settled by extended families as well as individuals. Often, several generations within a family would move west together to explore options for growth and new opportunities. (Photograph by D.W. Diamond.)

Three

A Community Develops

After the initial thrust of creating a settlement from a raw frontier, Craig citizens turned their attention to developing the town into a vigorous community that could serve the townspeople as well as the surrounding populace. By 1909, it was a certainty that the railroad would make it to Craig, and the optimism generated by that anticipated arrival energized the small town to push forward on many fronts.

The 1909 Enlarged Homestead Act increased the size of a homestead to 320 acres, encouraging a new wave of homesteaders into the region. Buoyed with the successes of the early years, several versions of a type of chamber of commerce formed (and ebbed), always with the idea of promoting the area to the outside world. By 1911, a second newspaper was established which operated for 18 years. That same year, political forces at the state level forced the split of Routt County into two counties. Moffat County was created from the western two-thirds of the original county, and a new courthouse was completed in 1918.

The arrival of the railroad in 1913 opened up the area to a time of expansion not seen before, and goods and passengers traveled the newly established east-west corridor between Craig and Denver. The installations of water and sewer systems and an electrical plant were completed in just a few years' time. Organizations—civic, religious, and fraternal—formed and became instrumental in the forward momentum of the town.

Even the onset of the Great War could not totally dampen the enthusiasm of the growing town. Patriotic young men, and a few women too, left to join in the overseas fight. Throughout the conflict, the small town continued to look forward as it spread its wings in optimistic development.

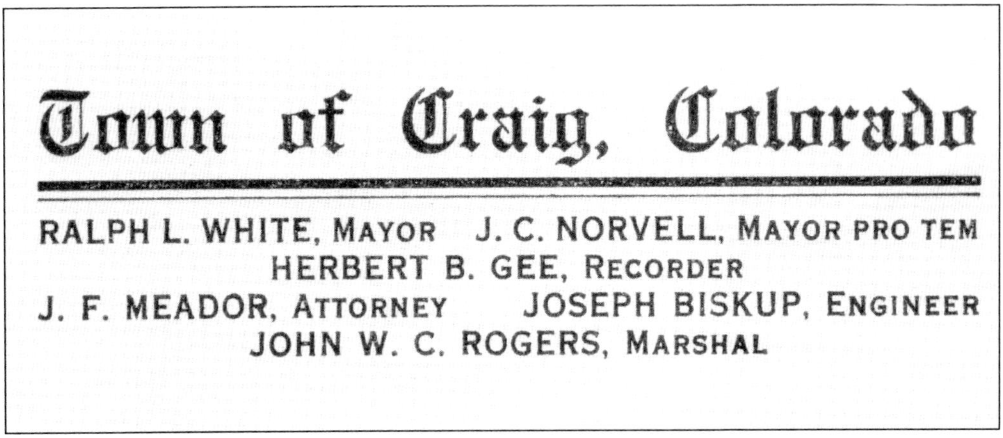
Town of Craig, Colorado

RALPH L. WHITE, MAYOR J. C. NORVELL, MAYOR PRO TEM
HERBERT B. GEE, RECORDER
J. F. MEADOR, ATTORNEY JOSEPH BISKUP, ENGINEER
JOHN W. C. ROGERS, MARSHAL

CRAIG LETTERHEAD, 1917. This is an example of letterhead for the Town of Craig.

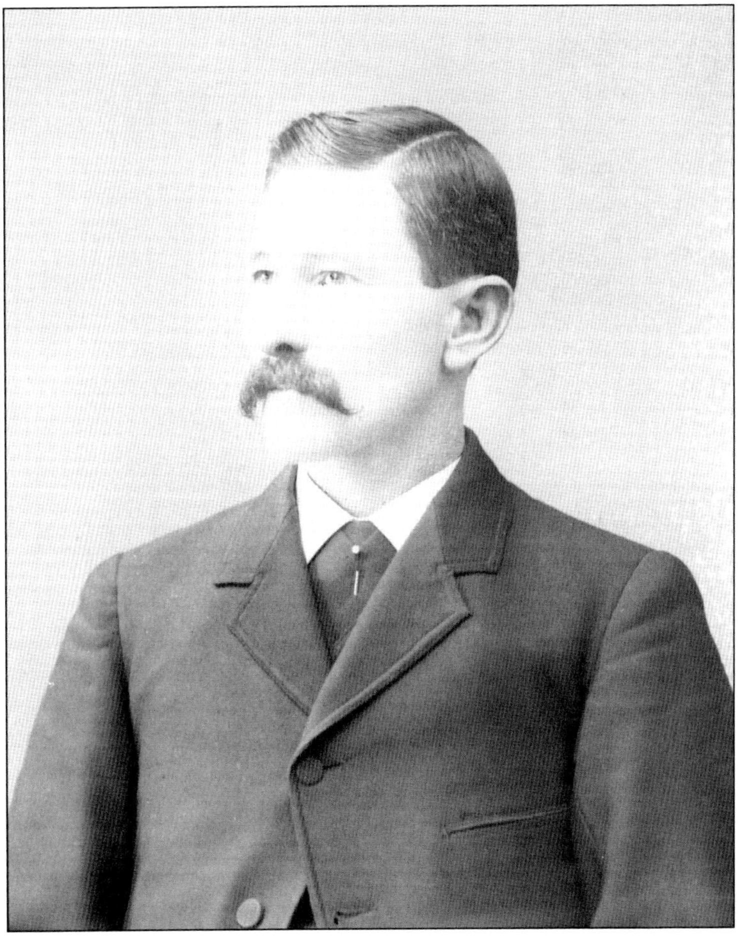

ALEXANDER S. ROBINSON. Alexander Robinson—half brother of Craig cofounder Willard Teagarden and brother-in-law of W.H. Tucker—moved to Craig with his wife, Annie, in 1890. Robinson was a carpenter and built many of Craig's early buildings. In 1908, he was elected Craig's first mayor when the little frontier town was incorporated. The Robinsons retired to Denver in 1919, after 30 years of contributing to the growth of the young community.

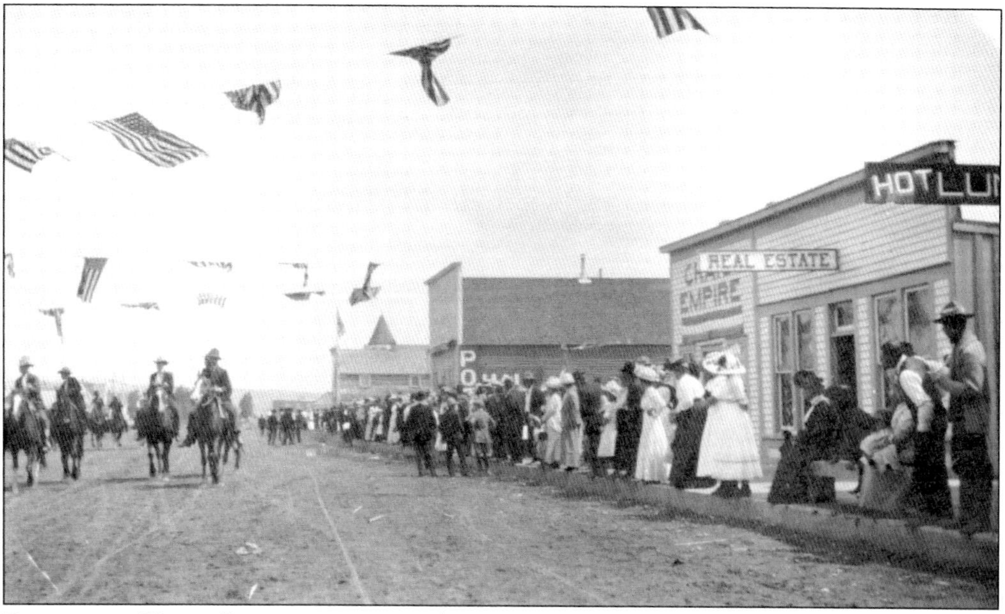

DOWNTOWN HORSE RACE, 1914. An unknown photographer caught the crowd waiting for a horse race to begin in downtown Craig. The view was taken in the 500 block of Yampa Avenue, looking north. The original Craig Empire building is on the right. As the town grew, community activities, such as this one, became more common.

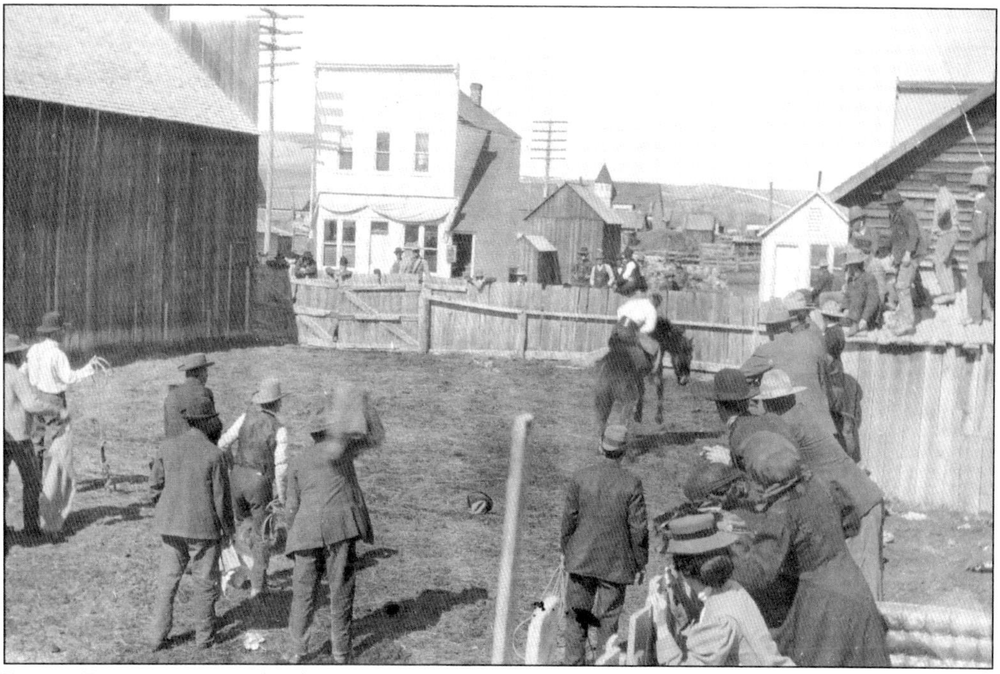

BRONC RIDING, 1910. This bronc ride took place in downtown Craig. The photograph shows the south side of the lot at 40 East Victory Way, across from the present-day West Theater. (Photograph by L.M. Hellebust.)

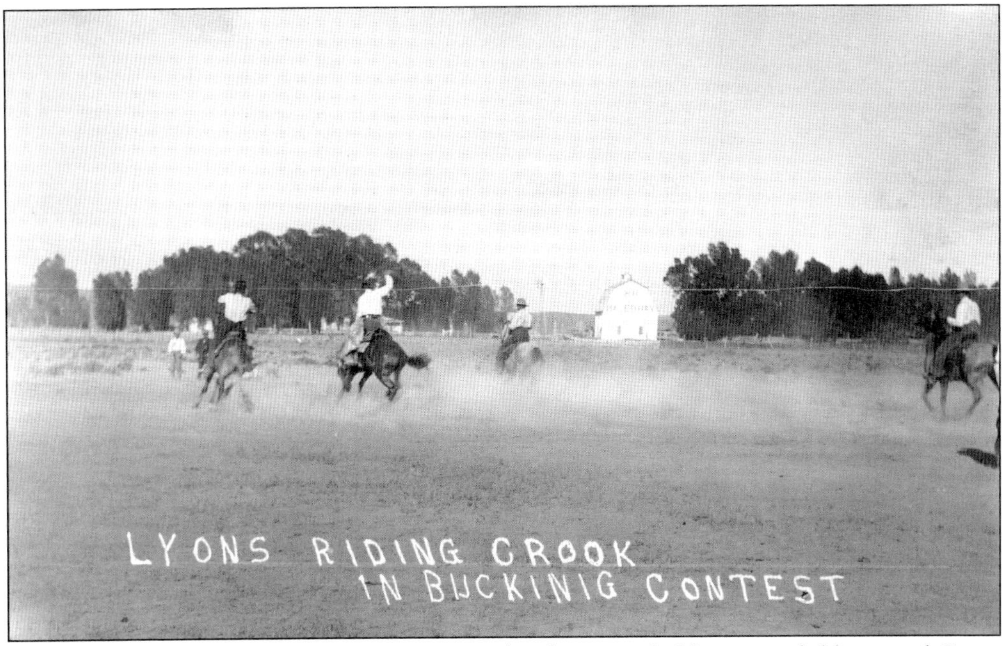

Bucking Contest, 1910. Bucking contests and rodeos were held in open fields around Craig. This photograph was taken near the William H. Rose barn by Fortification Creek. (Photograph by L.M. Hellebust.)

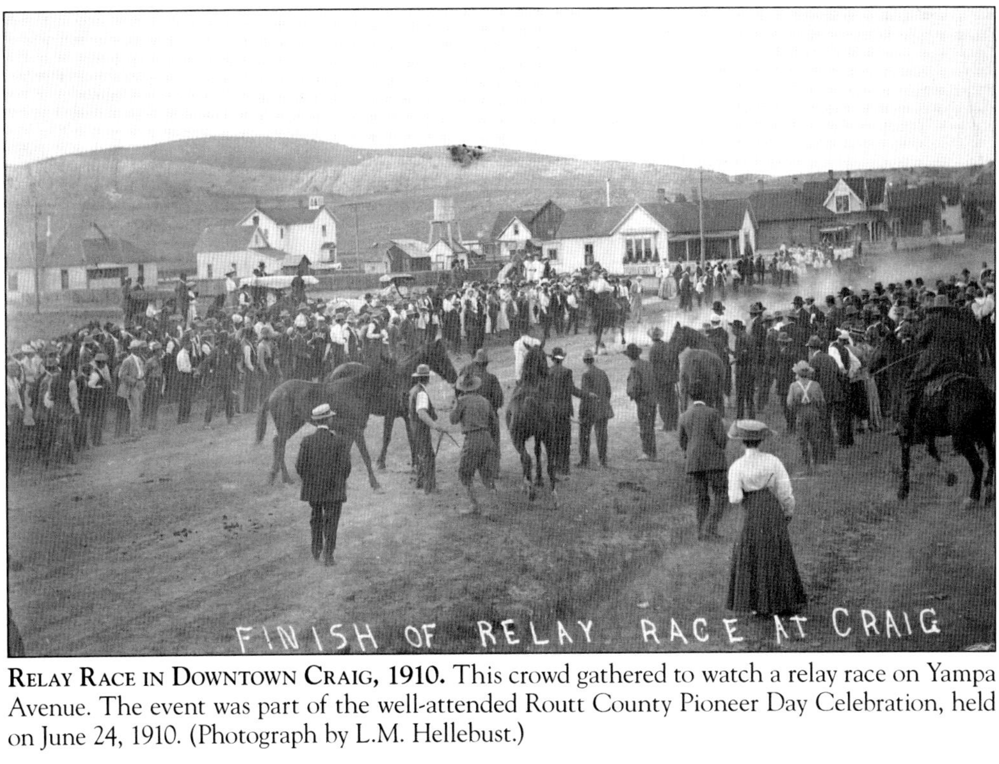

Relay Race in Downtown Craig, 1910. This crowd gathered to watch a relay race on Yampa Avenue. The event was part of the well-attended Routt County Pioneer Day Celebration, held on June 24, 1910. (Photograph by L.M. Hellebust.)

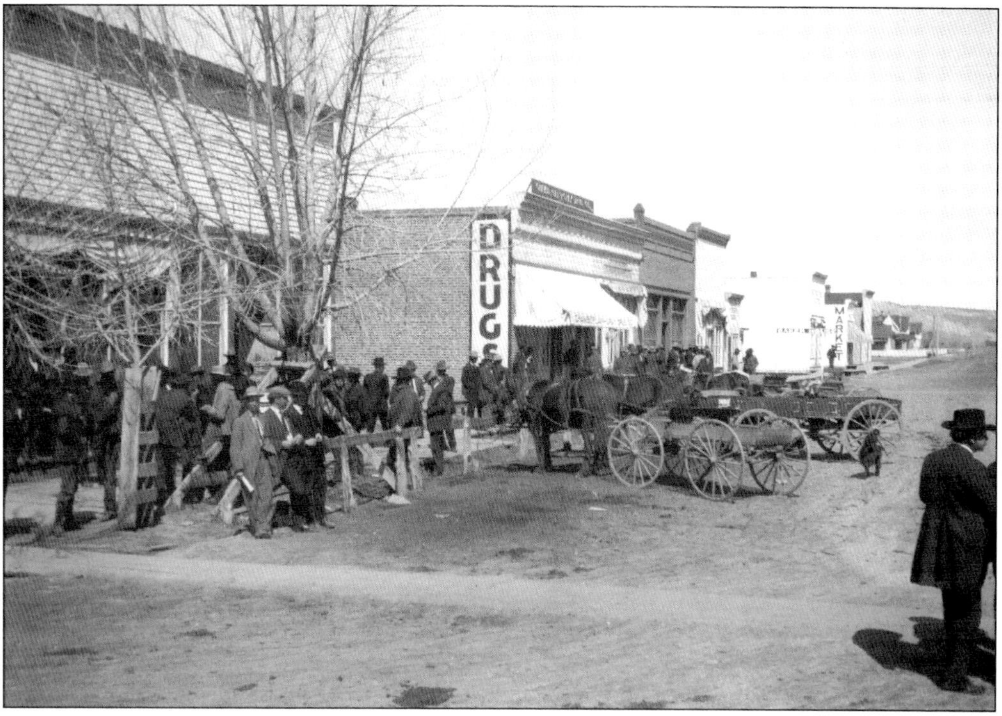

Chamberlain Gray Drug Store. Chamberlain Gray Drug Store was Craig's first chain drugstore. It opened in 1909 and was managed by Frank Cowan. In 1919, Cowan was able to purchase the store, which remained in business on Yampa Avenue for many years.

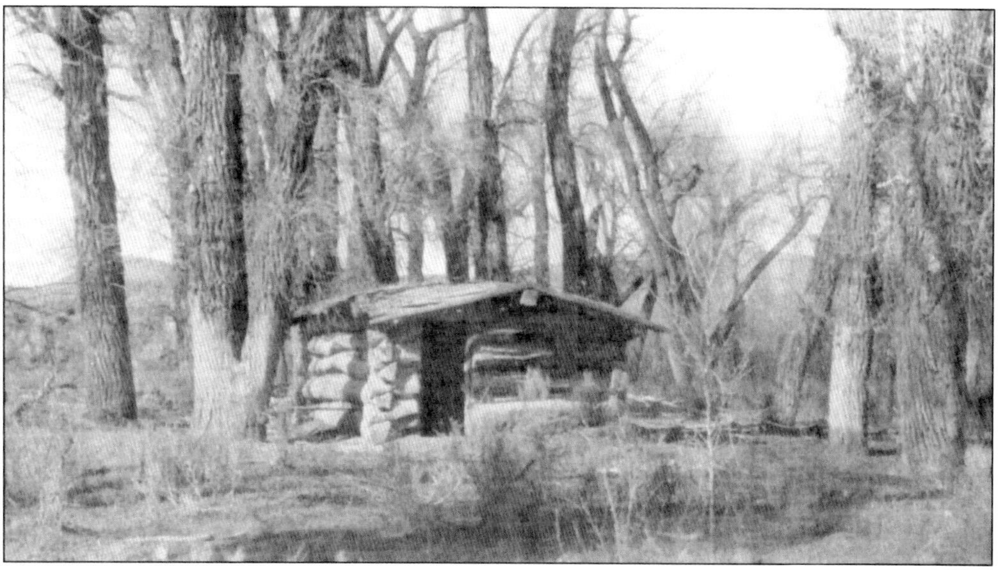

Pest House, 1909. This log structure was located near the river southwest of town and was used to quarantine those with smallpox. (Photograph by Robert M. Richardson IV.)

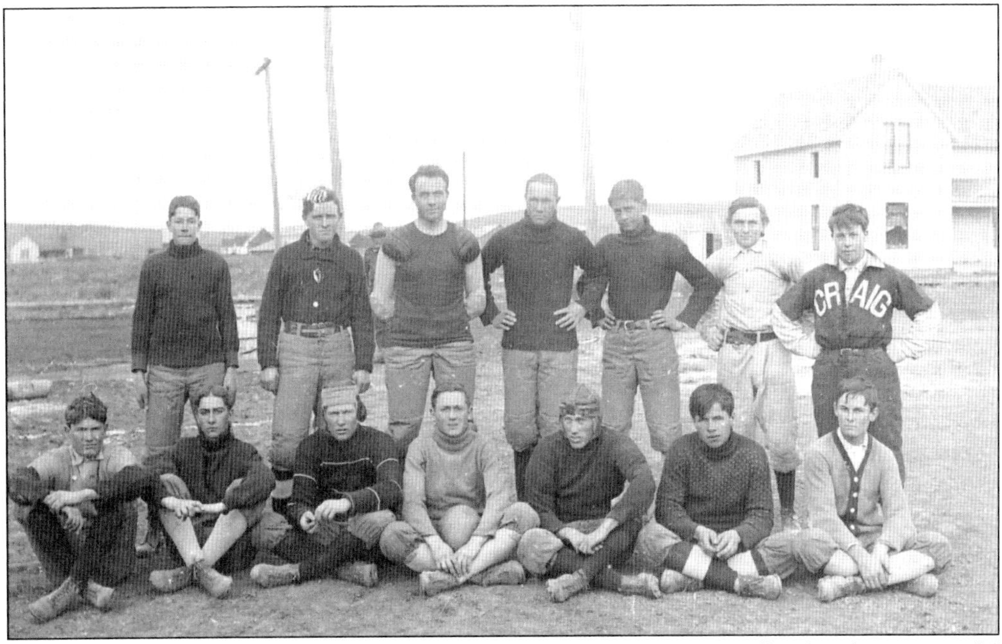

Champions. Members of the Craig High School football team pose proudly after their successful season, which gave them the valley championship for 1910. With continued growth in school enrollment, organized team sports became more numerous.

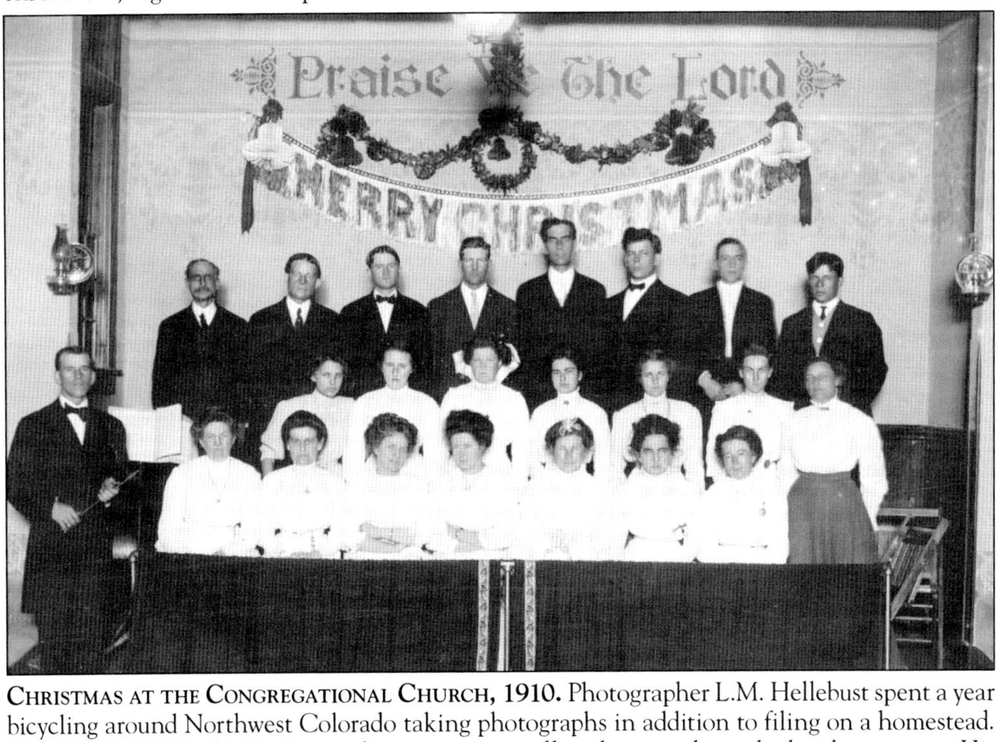

Christmas at the Congregational Church, 1910. Photographer L.M. Hellebust spent a year bicycling around Northwest Colorado taking photographs in addition to filing on a homestead. He left in 1911, but almost 50 years later, sent many of his photographs to the local newspaper. His clear, crisp images have been a gift to the community, with their splendid visual documentation of the growing town. (Photograph by L.M. Hellebust.)

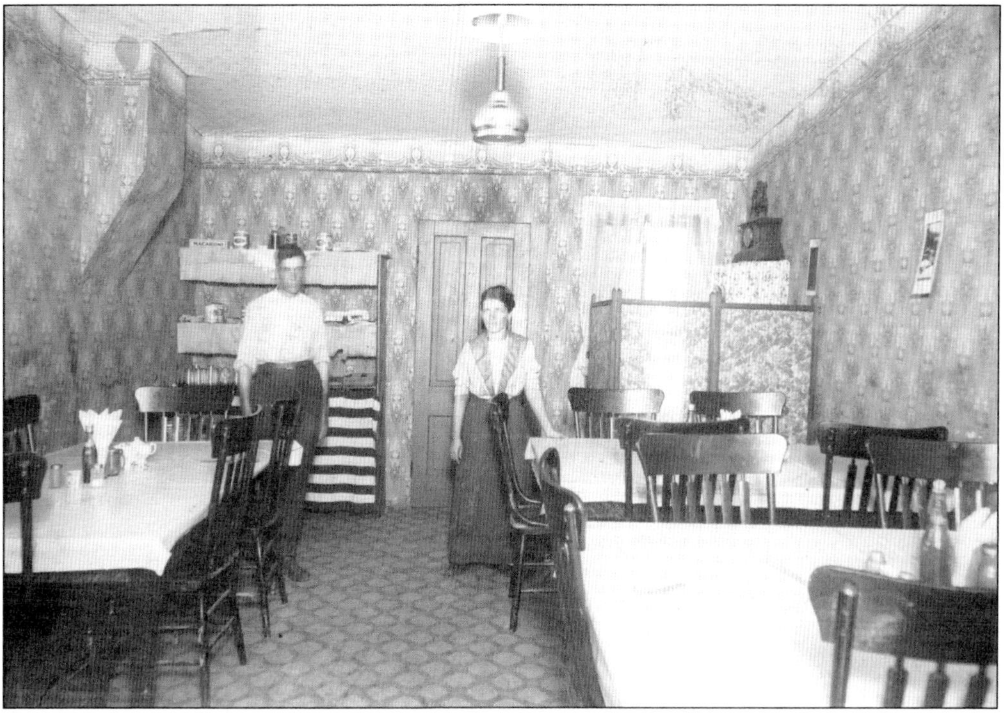

Craig Restaurant. Barbara Sommers, the proprietor of this Craig restaurant in 1911, poses with an unidentified man. The restaurant, one of several in Craig, was on Yampa Avenue.

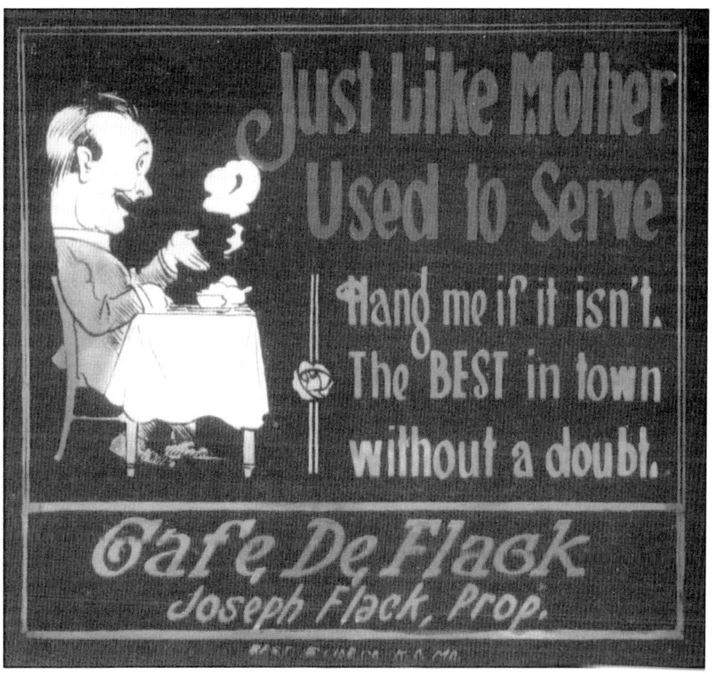

Café de Flack Advertisement. Joseph Flack was a retired and disabled Spanish American War veteran who moved to Craig in 1910. In 1912, he and his new wife opened a café on Yampa Avenue that he operated until 1915. Flack was a beloved citizen of the community, and in 1914, he was elected justice of the peace, a position he held until his death in 1925.

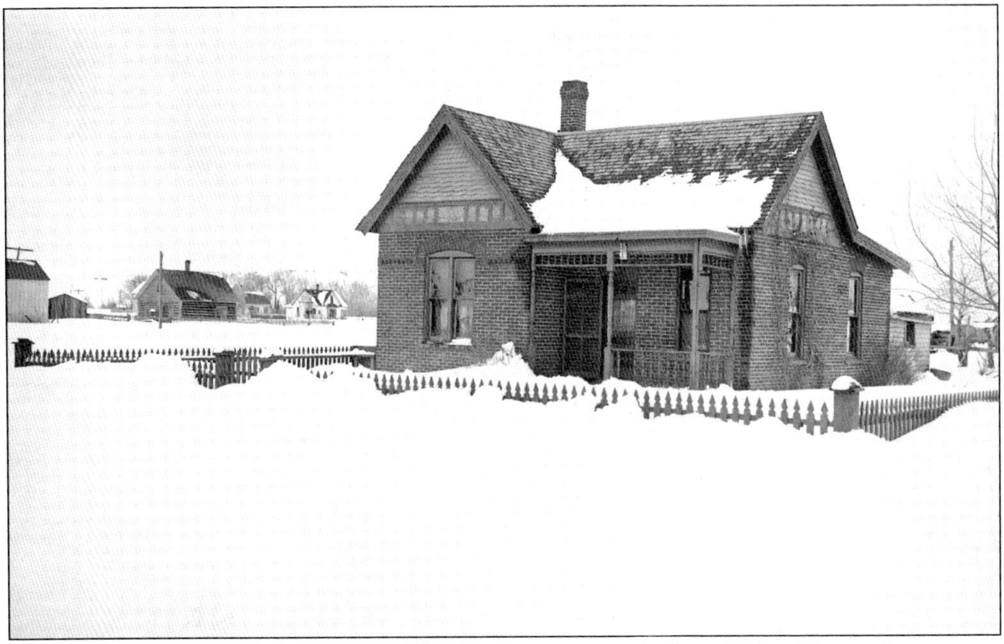

First Christian Church Parsonage. This brick cottage was built in 1895 on Yampa Avenue, just north of the church, for Rev. J.L. Ellis and his family. Later, it was leased by Mary W. Humphrey for a millinery shop after her husband died. In the 1910s, the house was purchased by the Drescher family and saw many changes over the decades before it was torn down in October 1988.

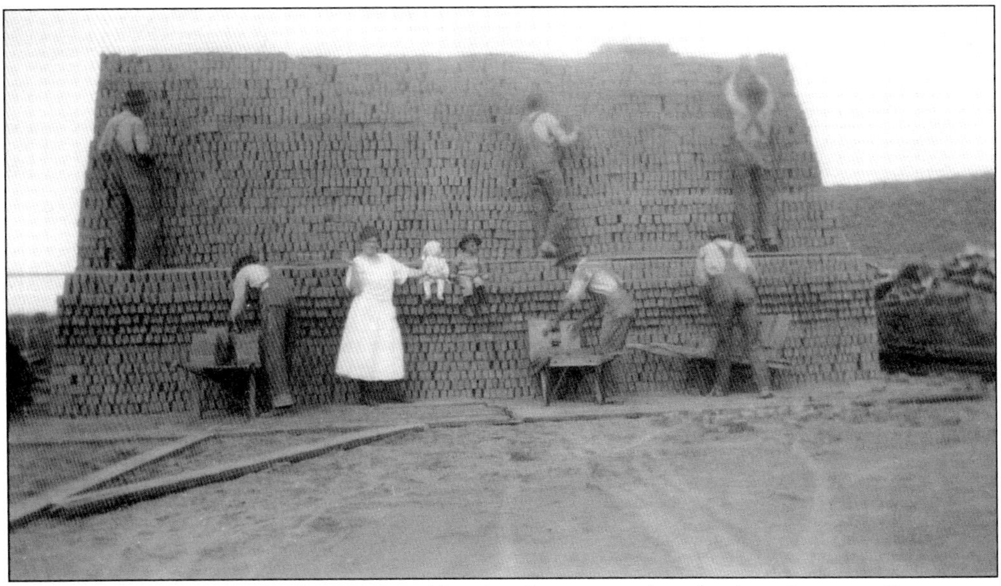

Big Employer, 1920. Charles Jones and his family pose in their brickyard in east Craig, near present-day Highway 40 and First Street. Starting off with an impressive production of 10,000 bricks per day, Jones planned to bring machinery that would boost production to twice that amount. He anticipated employing 12 people at the plant. Craig, during this time, was poised for a building boom, and Jones anticipated a large market for his products.

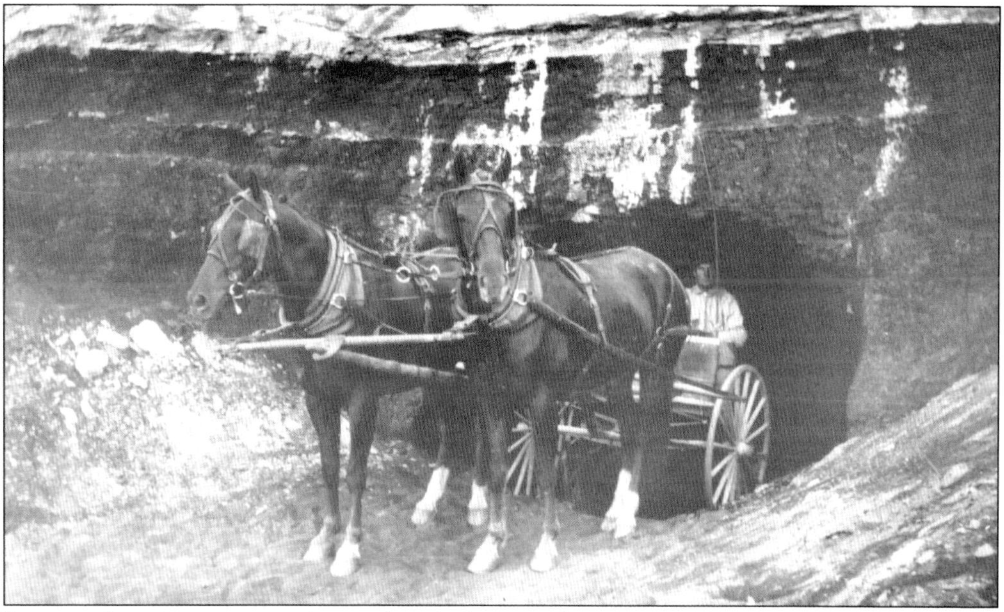

Blevins Coal Mine, 1912. An unidentified man pulls out of a coal seam at a mine near Craig. Coal, readily available and economical, was a main source of heat for many homes and businesses in the early days of Craig. (Photograph by E.T. Hancock; courtesy USGS.)

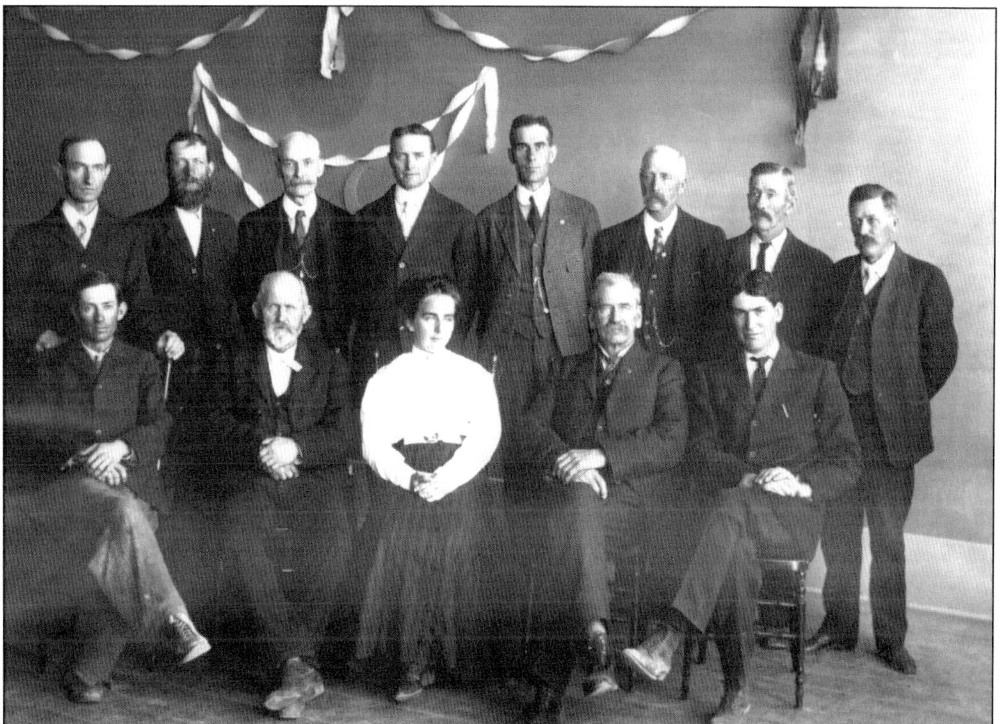

Promoting Craig, 1910. The Craig Commercial Association was a group of civic-minded individuals that promoted Craig and Moffat County. It was a forerunner of today's chamber of commerce.

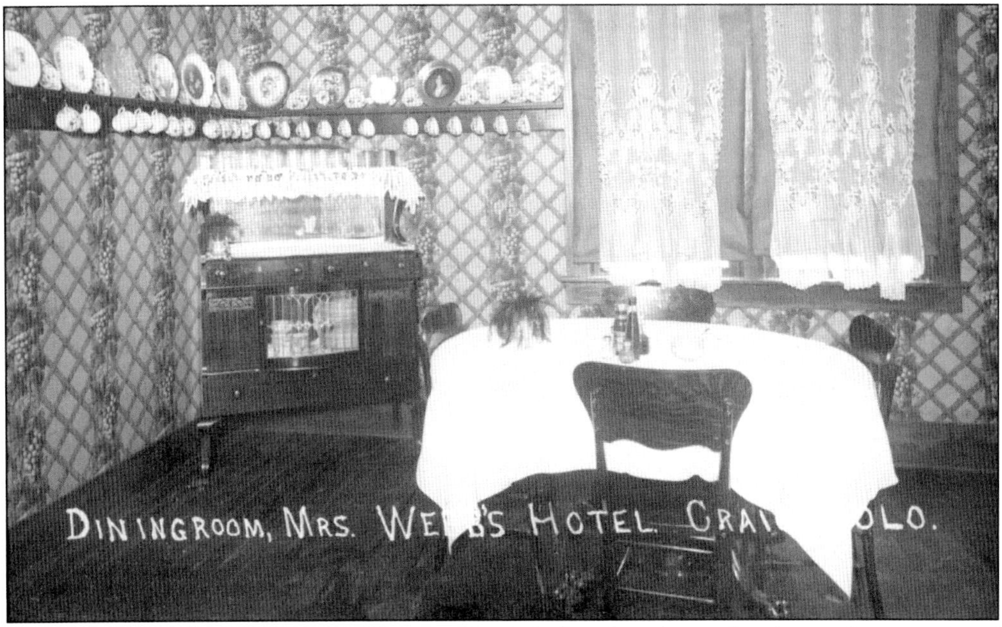

DINING ROOM AT ROSETTA WEBB'S HOTEL, 1910. The hotel was located at the northeast corner of Yampa Avenue and Main Street. Rosetta Webb, a beloved member of the community, was known for getting things accomplished around town. She organized the Craig Library Association in May 1911 and opened Craig's first library in the lobby of her hotel. The organization was legally incorporated in September 1912. Rosetta was also instrumental in the purchase of land for City Park. (Photograph by L.M. Hellebust.)

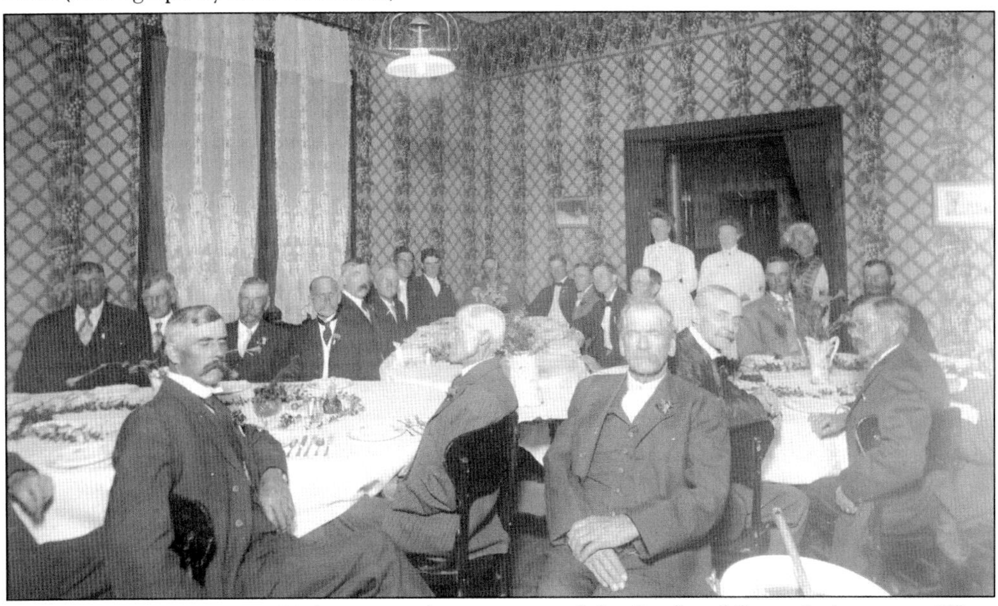

VETERANS DINNER, 1911. To bring together veterans of the Civil and Spanish-American Wars, Rosetta Webb hosted this special dinner in her hotel. The *Craig Empire* newspaper reported that unplanned excitement was caused when some of the party decorations caught on fire from a stray spark from Hellebust's "flashlight picture" of the guests. The fire was quickly extinguished. (Photograph by L.M. Hellebust.)

David Moffat, c. 1910. David Moffat, promoter of the direct rail connection between Denver and Salt Lake City, sits in his Denver office shortly before his death in 1911. Moffat poured his entire fortune and all his energy into building the rail line over the Rocky Mountains. After he died in New York City while seeking additional funding, the rails only made it into Craig, which is still the line's terminus today. (Photograph by L.C. McClure; courtesy Denver Public Library.)

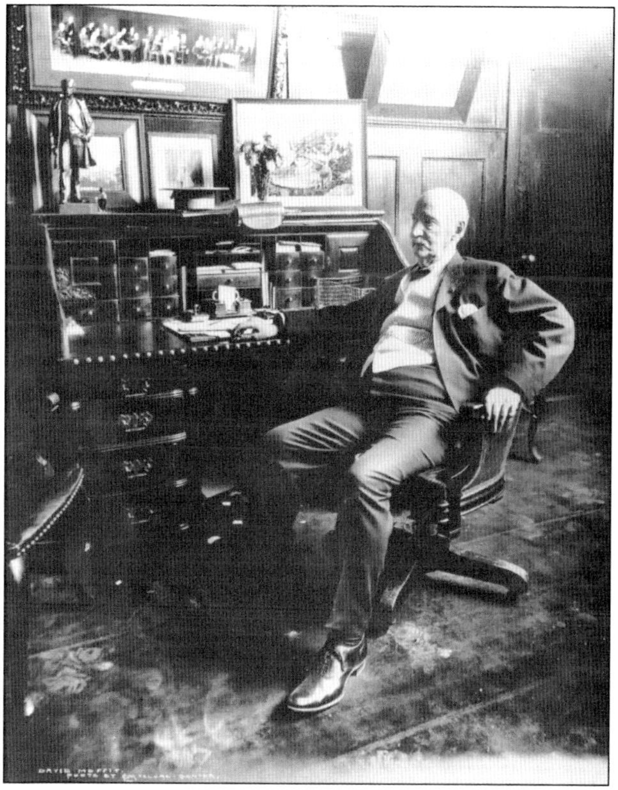

Moffat County Courthouse. The newly completed courthouse is pictured below in the spring of 1918. (Photograph by George Welch.)

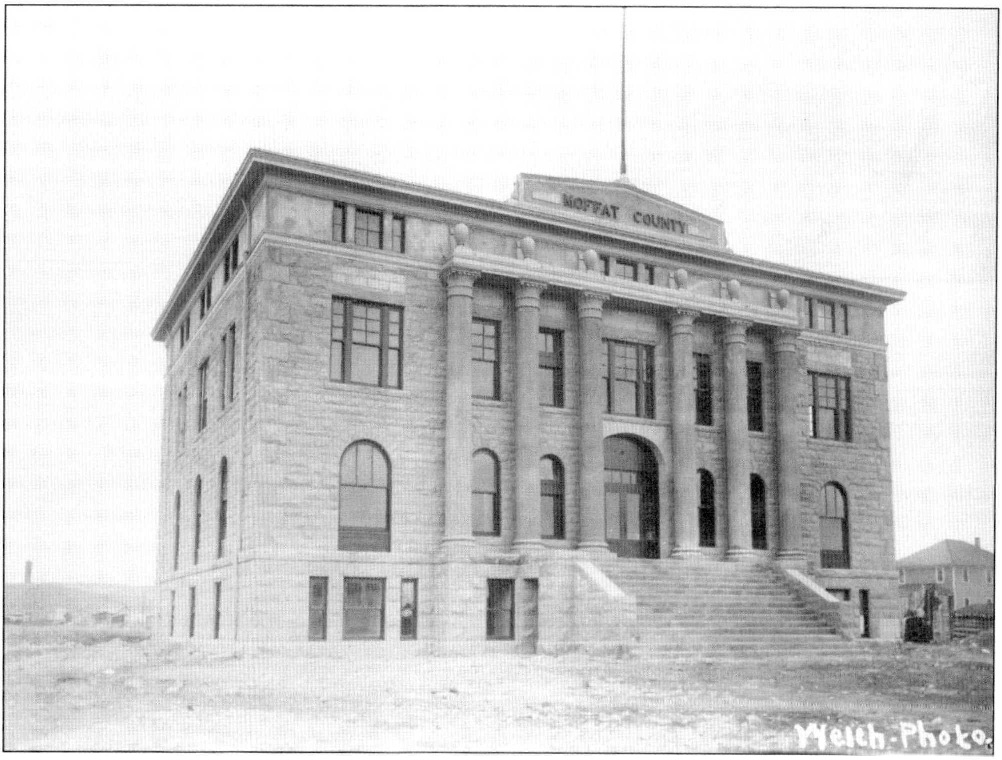

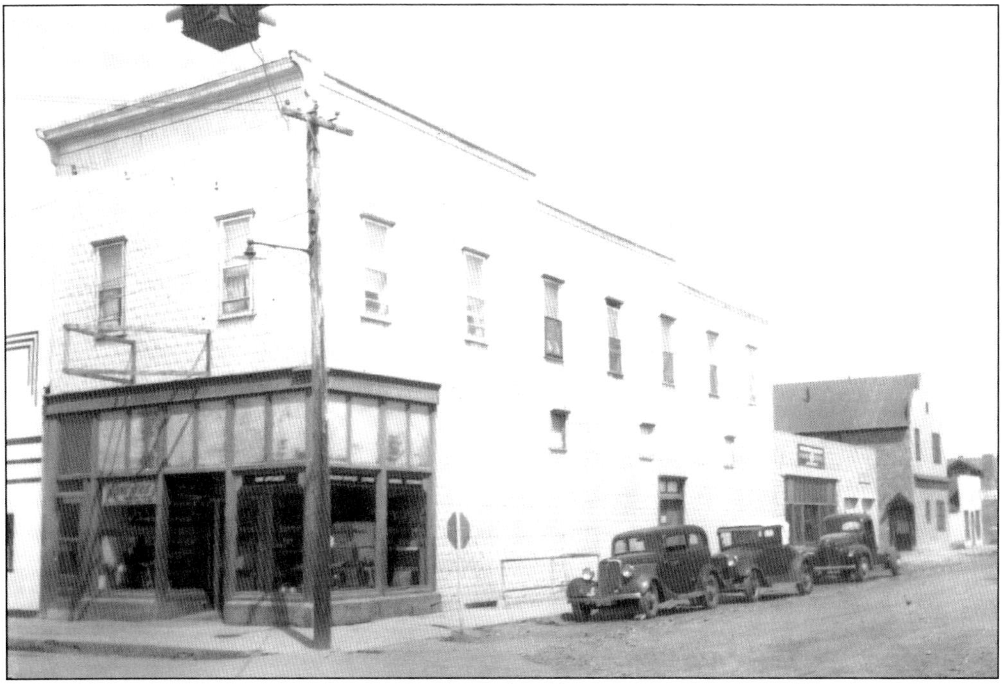

First Courthouse Building, c. 1925. Located at the northwest corner of West Victory Way and Breeze Streets, this building served as the courthouse for the newly designated Moffat County from 1911 to 1917, before the new courthouse was completed. The building burned in a spectacular fire in 1947, which took out several businesses. (Photograph by George Welch.)

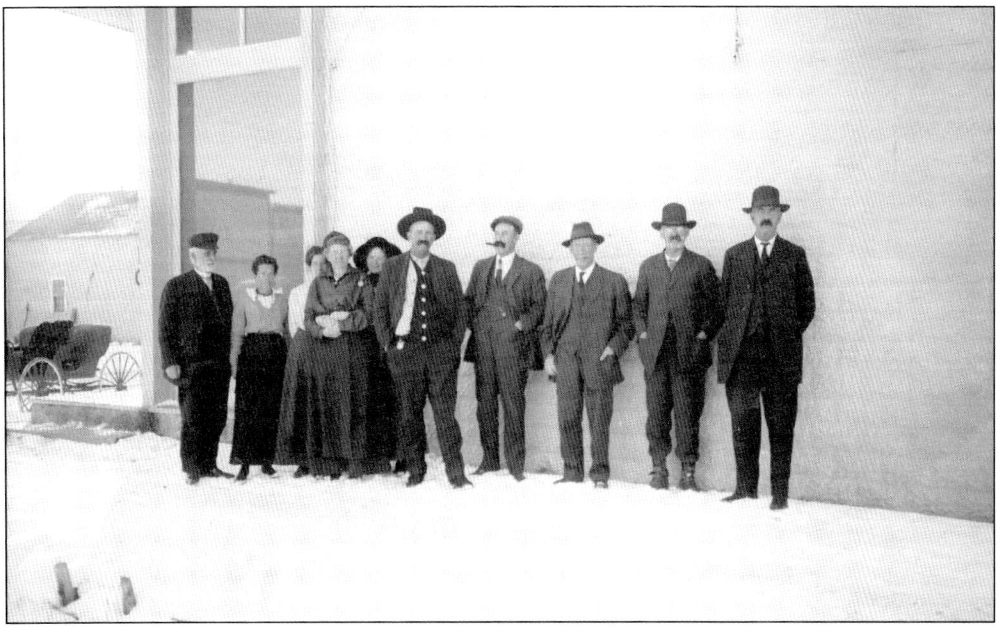

Moffat County Officials, 1914. Moffat County was formed from the western two-thirds of Routt County on February 27, 1911, with Craig as the county seat. In this photograph, early county officials pose with some of their assistants outside the building that served as the courthouse until one was built in 1917.

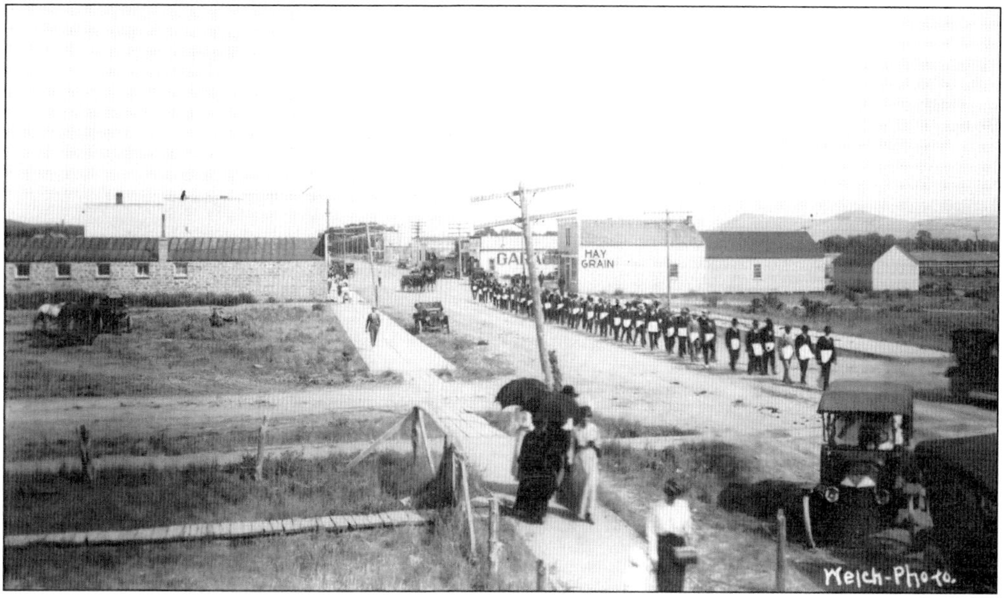

PARADE IN CRAIG, 1917. The Masons were the first fraternal organization in town. Here, they parade west on Main Street (now Victory Way) to the site of the new Moffat County Courthouse to celebrate the laying of the building's cornerstone on September 3. (Photograph by George Welch.)

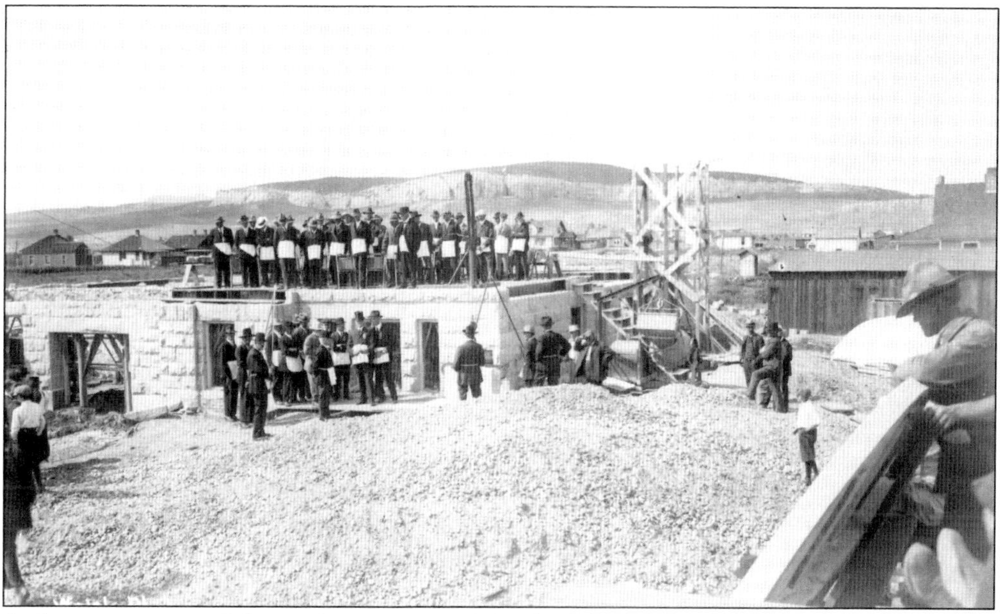

A CORNERSTONE FOR MOFFAT COUNTY, 1917. Men from the Masonic Lodge and other bystanders gathered for the laying of the cornerstone for the Moffat County Courthouse on September 3, 1917. Moffat County was formed in 1911, but it was six years before the county was able to build a proper courthouse.

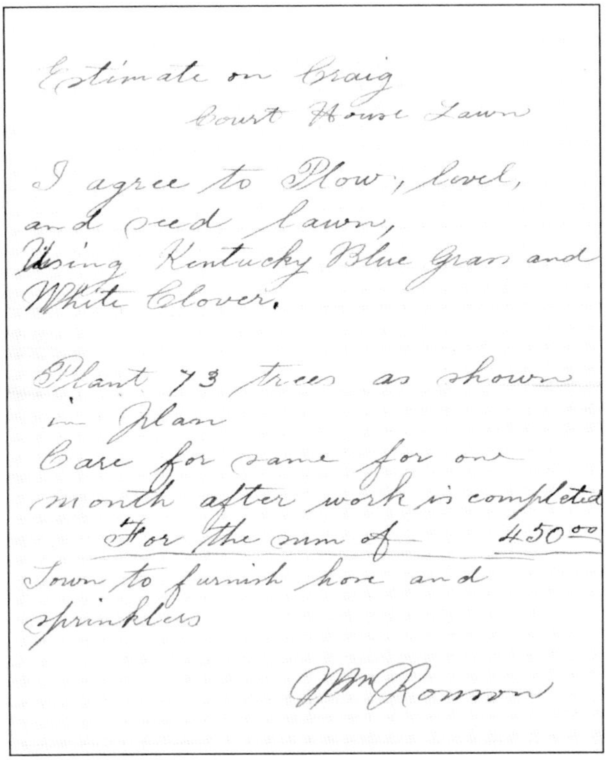
Estimate on Craig
Court House Lawn

I agree to Plow, level,
and seed lawn,
Using Kentucky Blue Grass and
White Clover.

Plant 73 trees as shown
in plan
Care for same for one
month after work is completed
For the sum of 450.00
Town to furnish hose and
sprinklers

Wm Rowan

MOFFAT COUNTY COURTHOUSE ESTIMATE, 1918. A $450 bid from William Rowan to landscape the courthouse lawn and plant trees included one month's care, but the town had to furnish the hose and sprinklers.

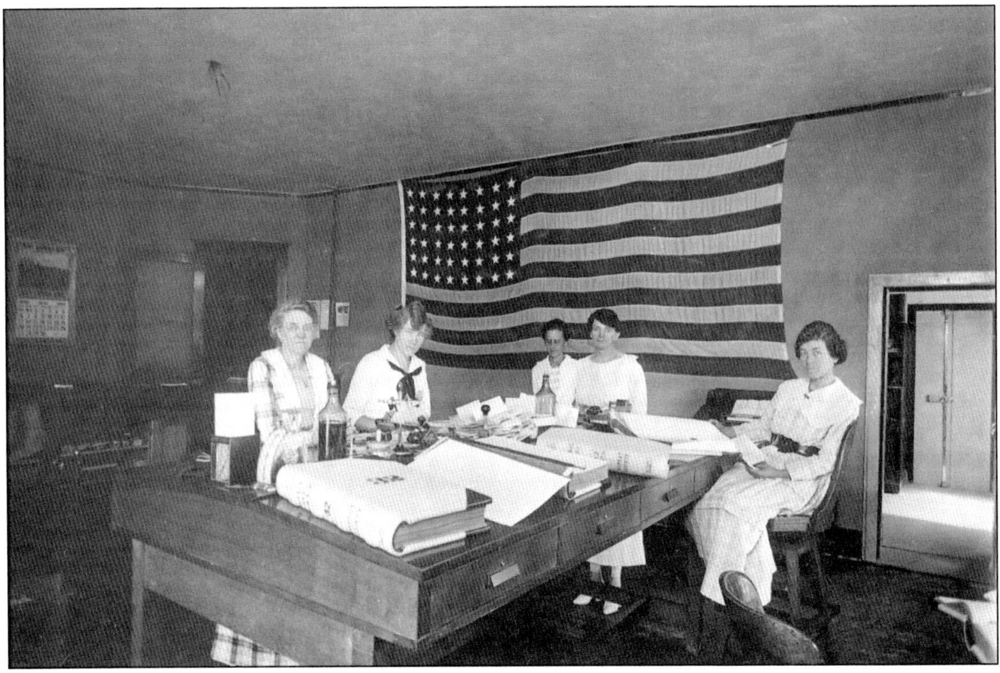

MOFFAT COUNTY CLERK AND RECORDER'S OFFICE, JUNE 1920. The new courthouse building was just three years old at that time of this photograph. The women in the recorder's office are, from left to right, Lillie Haughey (clerk and recorder), Maude Cooper, Julia Carpenter, Mollie Gregory, and Ossa Cooper. (Photograph by George Welch.)

HARRY DURHAM AND OLIVE VAN TASSEL, C. 1911. The Durhams were the first couple to be married in Moffat County whose place of residence (Moffat County) was listed on their license. Harry was born in Meeker in 1888, and Olive came to Breeze Basin with her parents in 1903. The couple lived on their ranch near Waddle Creek, where they raised seven children.

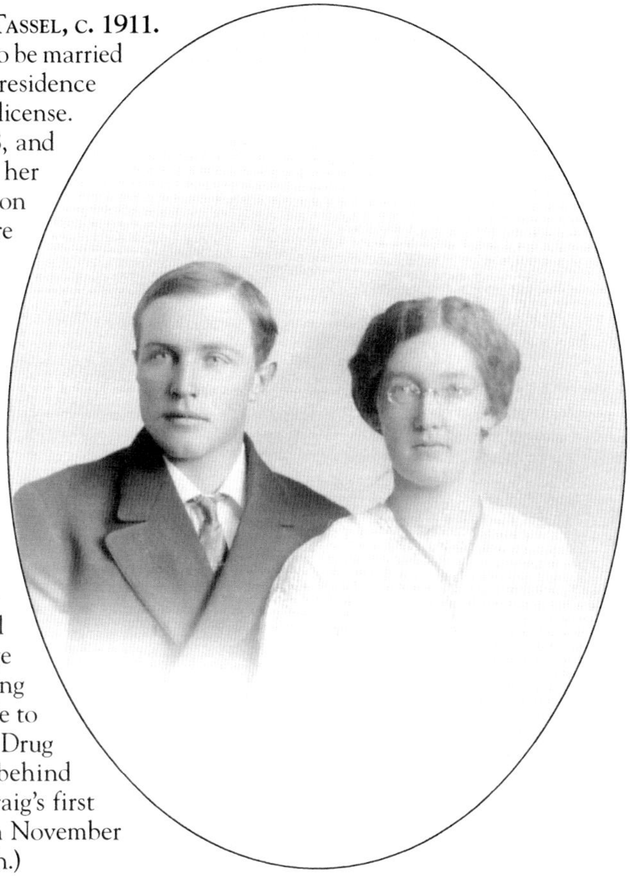

CULVERT INSTALLATION, 1915. This truck, pictured on Yampa Avenue across from the First National Bank building, was carrying a large drainage culvert on its bed, indicating that improvements were being made to local roads. The Chamberlain Gray Drug Store is on the right. The rubble behind the truck bed is the remains of Craig's first business building, which burned in November 1914. (Photograph by George Welch.)

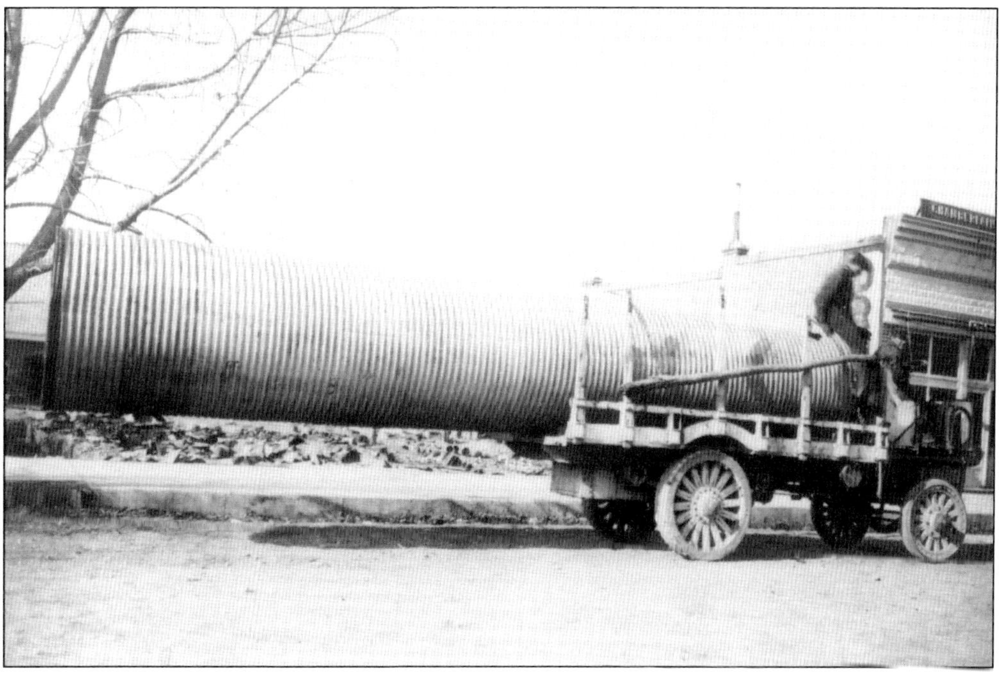

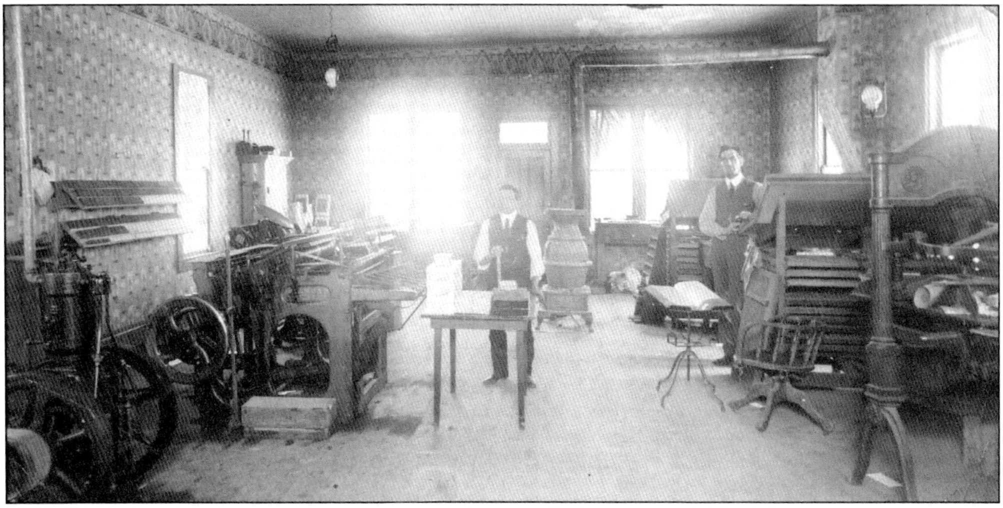

The Craig Courier Newspaper Office, 1915. This photograph of the interior of the *Craig Courier* newspaper office shows editor Walter Spencer seated at his desk. This newspaper, called the *Pantagraph* by its first owner, was started in 1891. The *Craig Courier* was in business until its merger with the *Craig Empire* in the late 1920s.

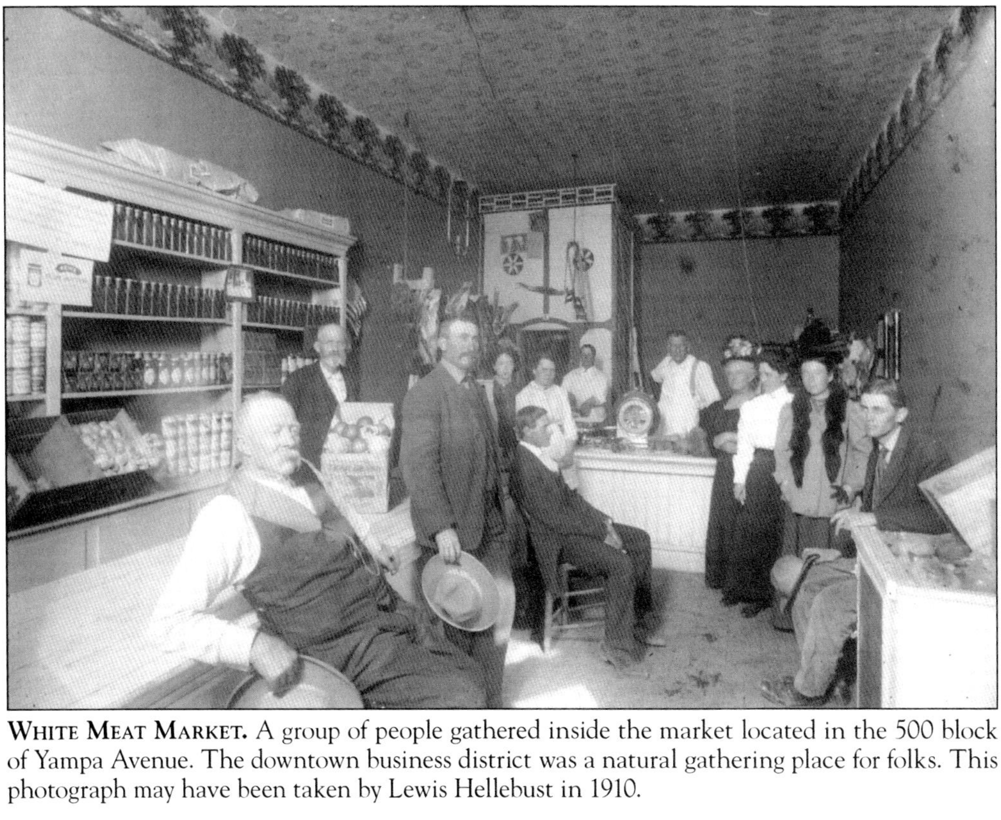

White Meat Market. A group of people gathered inside the market located in the 500 block of Yampa Avenue. The downtown business district was a natural gathering place for folks. This photograph may have been taken by Lewis Hellebust in 1910.

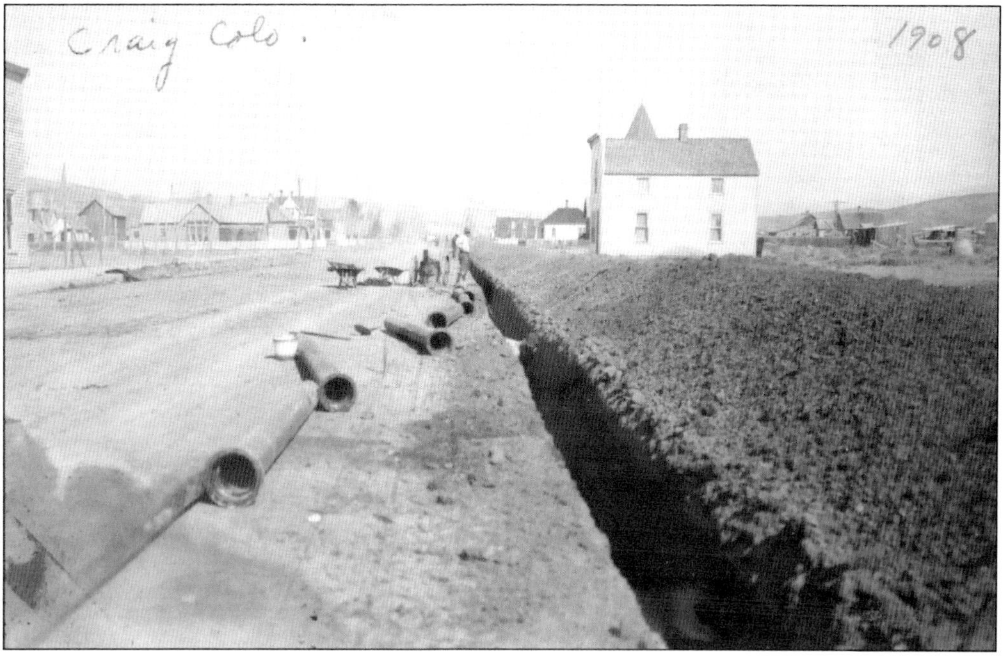

Laying Water Pipes, 1915. Men worked to lay pipes for the new water system, which was installed in Craig that year. After several disastrous fires, a water system for fire prevention was a priority for the town. The following year, a sewer system was installed. This view is looking north on Yampa Avenue.

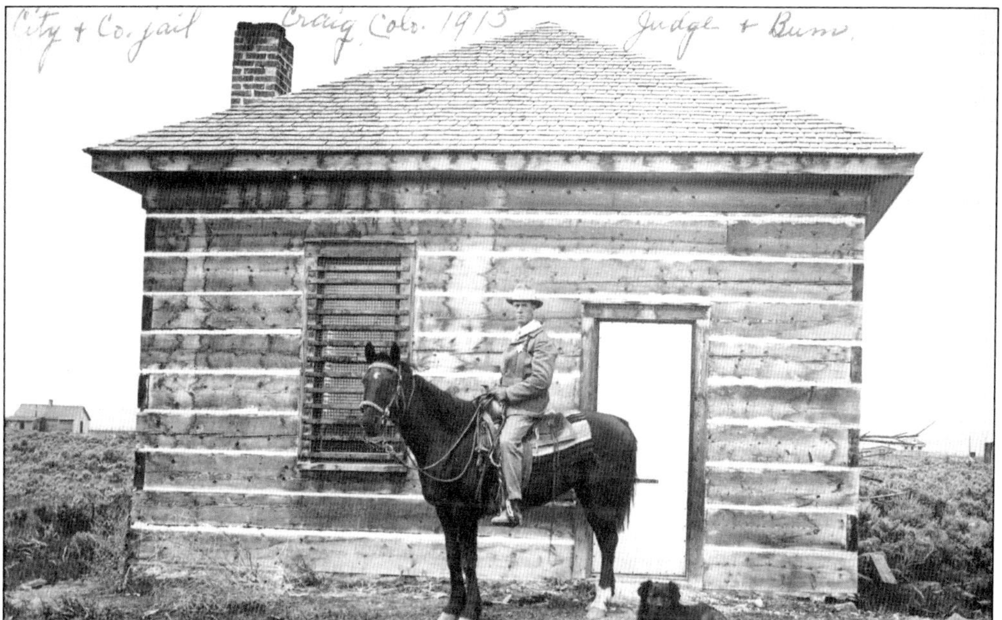

Craig's First Jail, 1915. Early town marshal John W.C. Rogers posed in front of the jail, located on the west side of the 400 block of Breeze Street. The jail was replaced when the Moffat County Courthouse was completed in 1917. (Photograph by George Welch.)

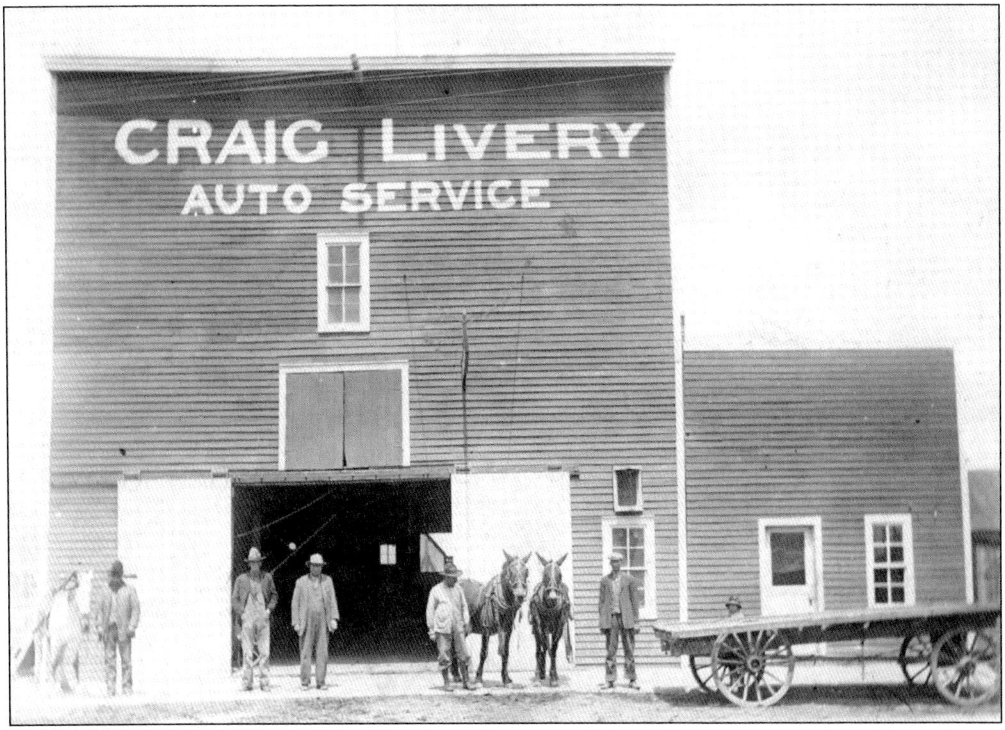

ON THE CUTTING EDGE, C. 1915. The Craig Livery and Auto Service, located on Main Street just west of Yampa Avenue, was one of the first liveries in town. The owners quickly adapted to the arrival of the first cars and added automobile repairs to their services.

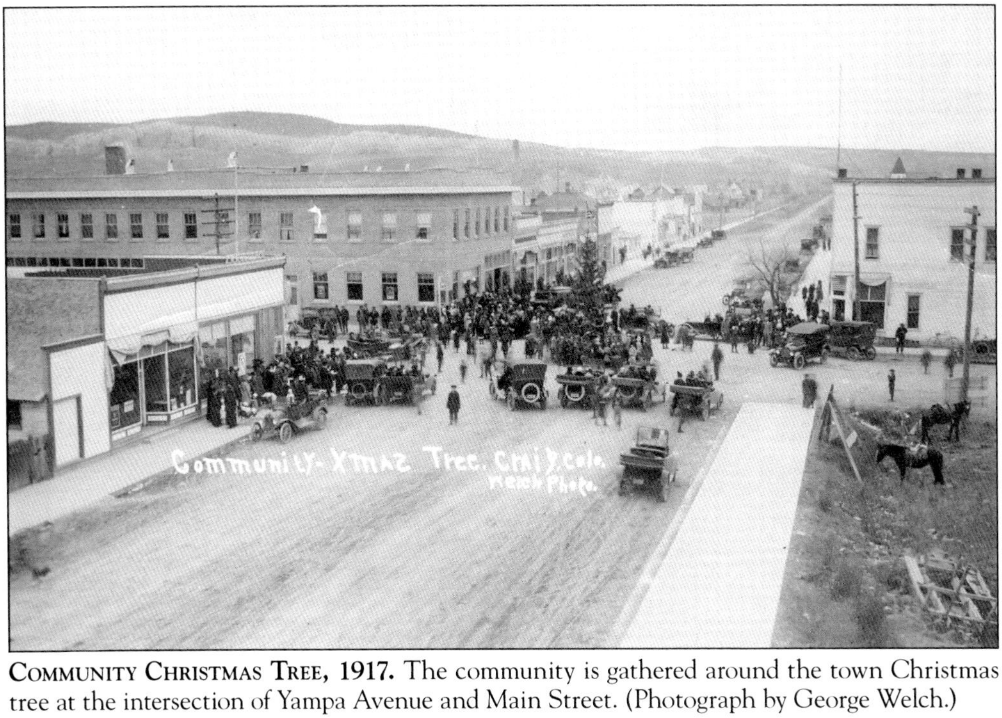

COMMUNITY CHRISTMAS TREE, 1917. The community is gathered around the town Christmas tree at the intersection of Yampa Avenue and Main Street. (Photograph by George Welch.)

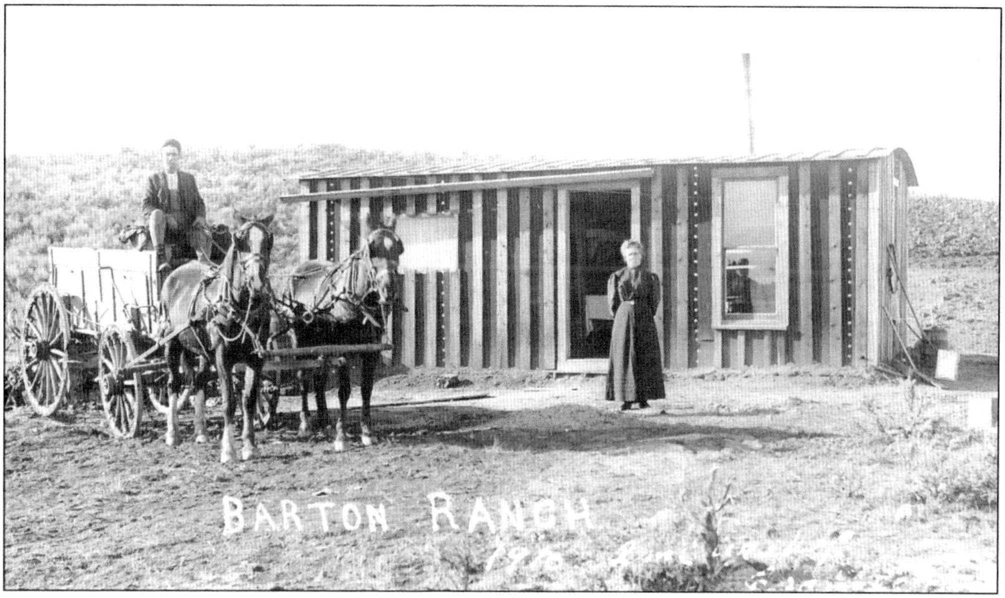

BARTON HOMESTEAD, 1910. Etta Barton and her son Allen posed outside Etta's homestead cabin. She located two miles west of Craig but did not stay long in the area. The Bartons were part of a large new wave of homesteaders who came into the area after the 1909 Homestead Enlargement Act. (Photograph by L.M. Hellebust.)

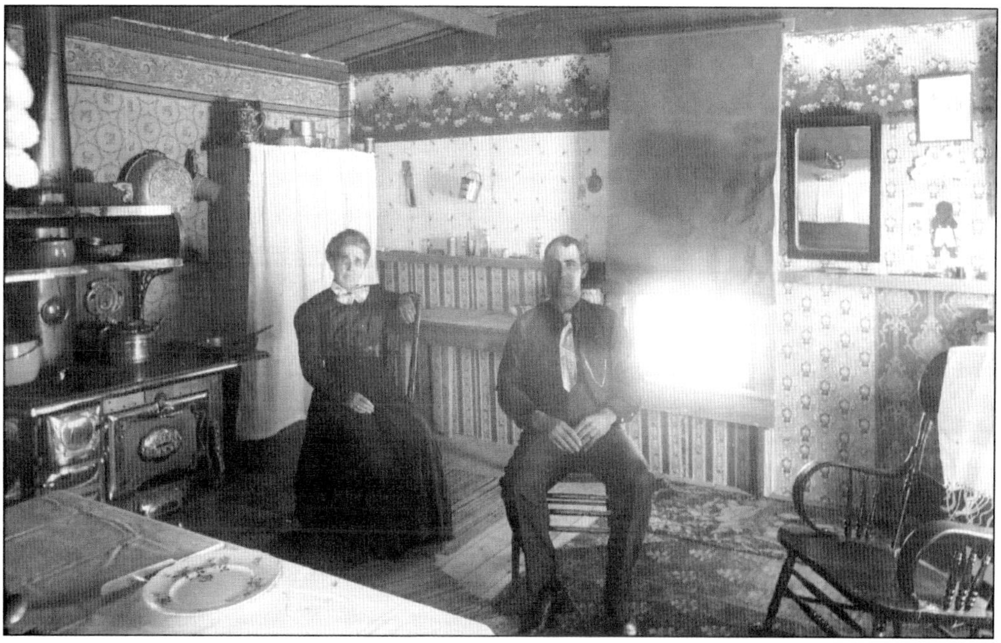

HOMESTEAD INTERIOR. Etta Barton homesteaded two miles west of Craig to be near her son who was living in the area. Etta's small house, complete with curtains, wallpaper, and throw rugs, was a far cry from the rough living quarters of the early pioneer homes. By 1920, both Etta and her son had left the area. (Photograph by L.M. Hellebust.)

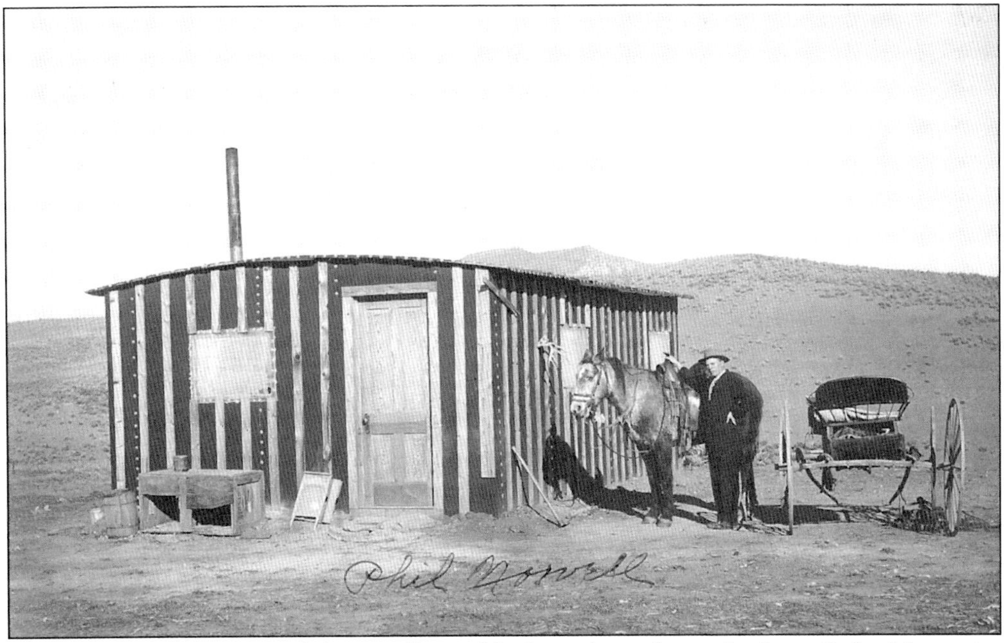

HOMESTEAD NEAR CRAIG. Phil Norvell posed by his cabin a few miles from Cedar Mountain. Many homesteader cabins of this era closely resembled one another. They appeared to have been built to barely meet the sparse specifications required by the Homestead Act. (Photograph by Lewis M. Hellebust.)

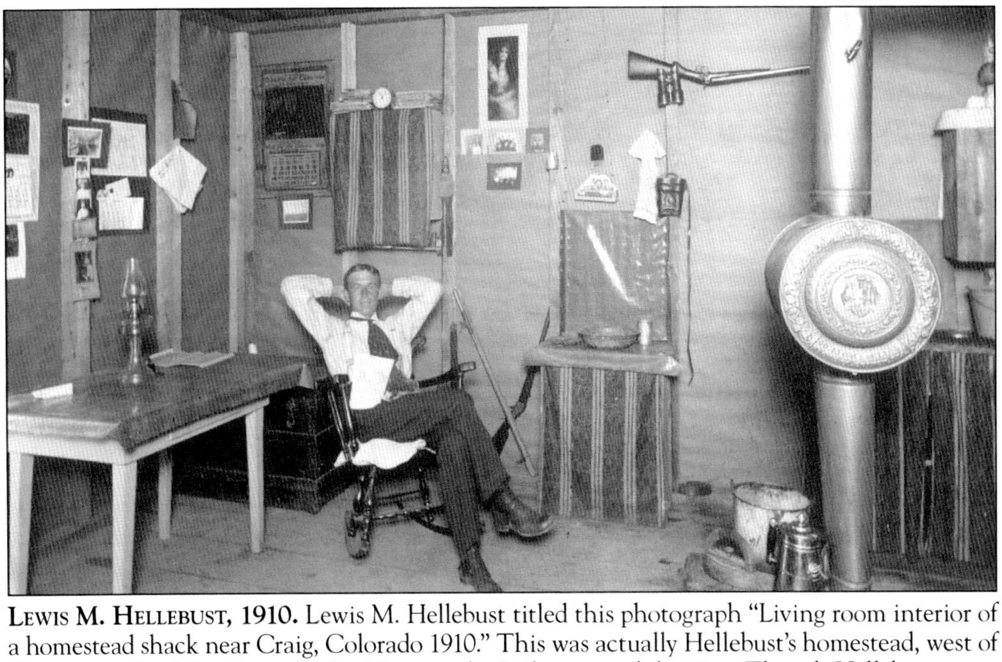

LEWIS M. HELLEBUST, 1910. Lewis M. Hellebust titled this photograph "Living room interior of a homestead shack near Craig, Colorado 1910." This was actually Hellebust's homestead, west of the present-day John Deere dealership near the Ridgeview subdivision. Though Hellebust never proved-up on his homestead, he took many fine images of the growing community during the year he was here. (Photograph by Lewis M. Hellebust.)

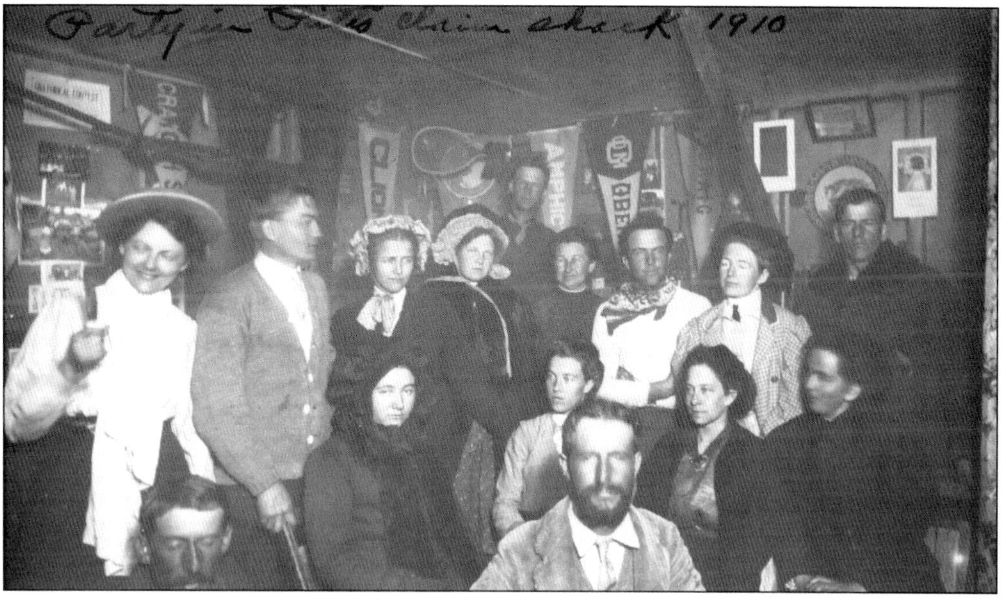

Partying at Pitt's Place. Party-goers are pictured at the G. Paul Pitt homestead just west of Craig. The homesteaders seemed, for the most part, to be socially inclined, and gatherings were frequent. Note the Craig high school pennant on the back wall and a calendar with the page for January 1911 hanging nearby. (Photograph by L.M. Hellebust.)

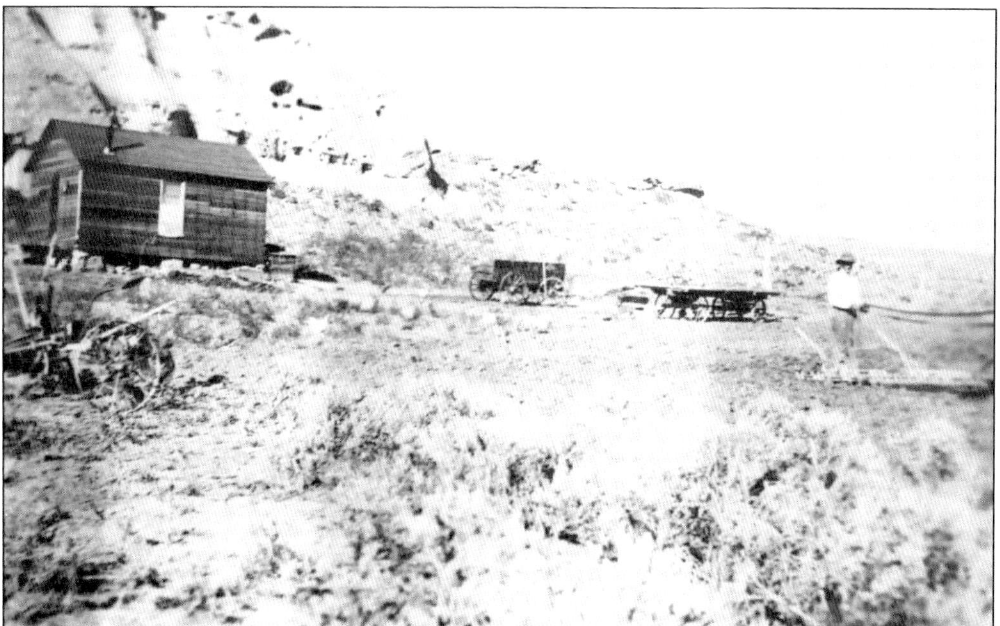

Carl and Myrtle Van Dorn Homestead, 1918. Myrtle Bryan moved to Northwest Colorado in 1885, at age two, with her parents Robert and Lucy Bryan. Lucy died shortly after, leaving Robert to raise Myrtle and her sister Maud. Myrtle married Carl, who had come to the Craig area in 1904, and they homesteaded northeast of Craig, in the present-day Glen Erie subdivision.

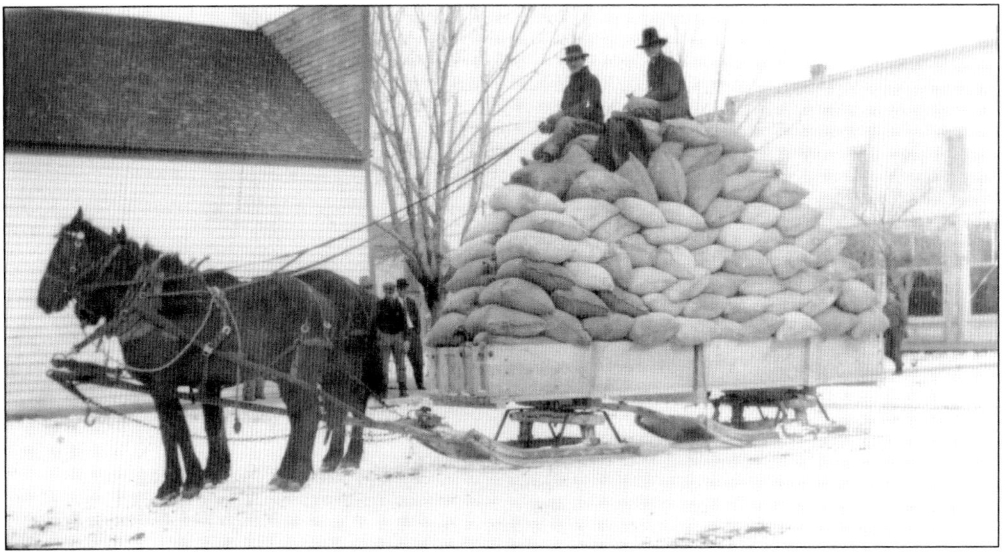

WAGON LOAD OF GRAIN, 1910. Two men perch atop a load of grain in downtown Craig at the corner of Main Street (now Victory Way) and Yampa Avenue. The first commercial building in Craig is on the left, which was built to serve the Craig Land and Mercantile Company as a land sales office, a mercantile, and a post office when the townsite was first platted. It burned in November 1914.

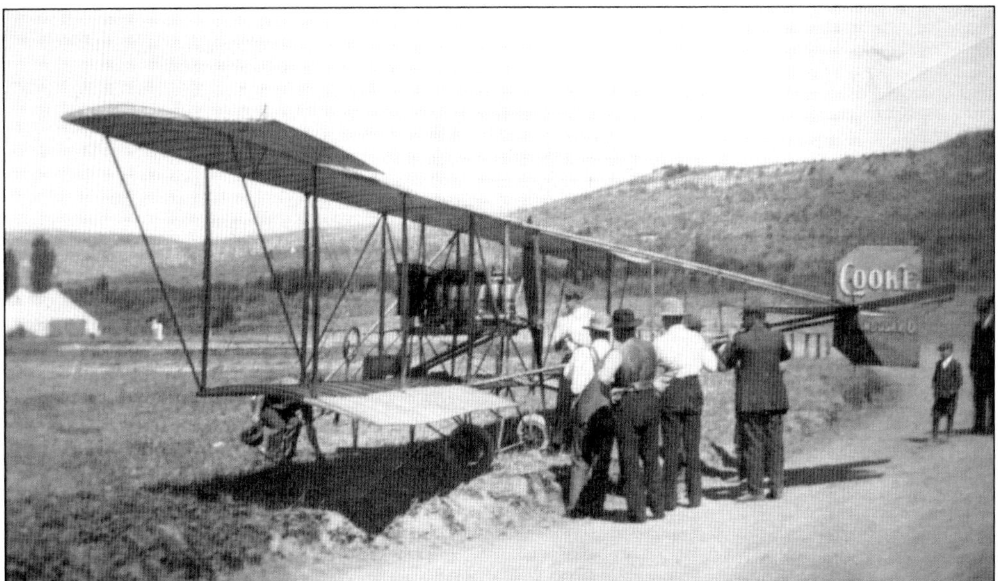

FIRST PLANE INTO THE YAMPA VALLEY. Weldon B. Cooke flew this biplane at the first Routt County Fair at Hayden, Colorado, in September 1914. Cooke was killed in a plane accident in Pueblo, Colorado, two weeks later. (Photograph by George Welch.)

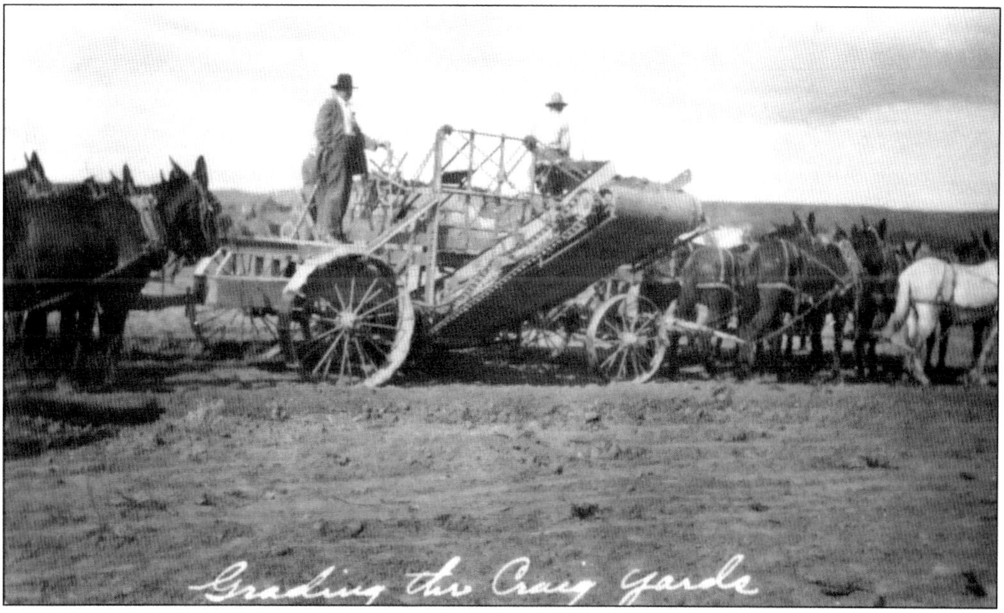

GETTING READY FOR THE RAILS, 1913. In 1913, the region was buzzing with excitement in anticipation of the arrival of the railroad. This image, titled "Grading the Craig Yards," shows a crew working to prepare the rail yard in Craig at the south end of Yampa Avenue.

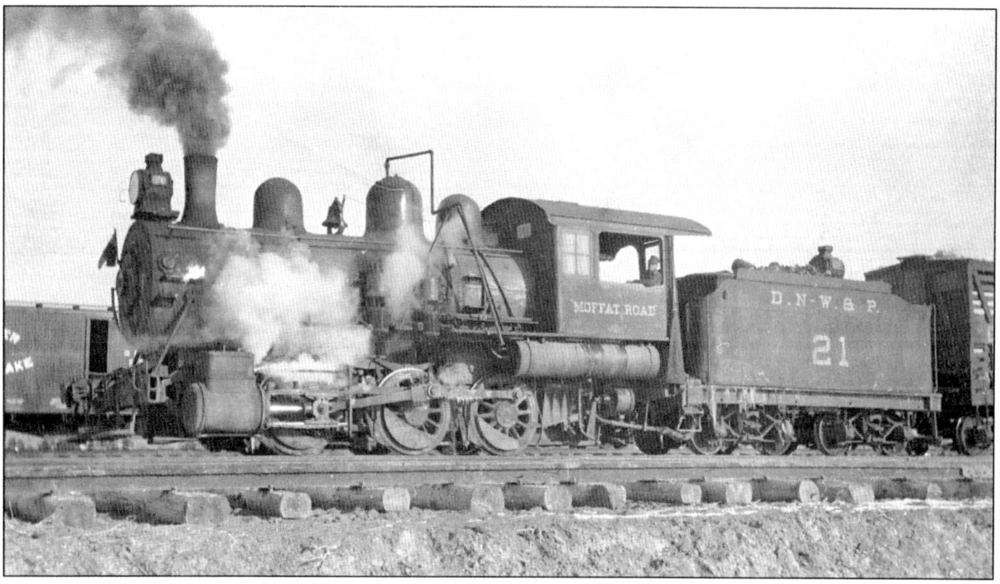

MOFFAT ROAD ENGINE. Engine 21 was used during much of the railroad construction while the line into Craig was being built in 1913. The coming of the railroad to Northwest Colorado was a momentous occasion and allowed area residents easier access to outside goods, commerce and travel, as well as providing a reliable way to transport local commodities to the other side of the mountains.

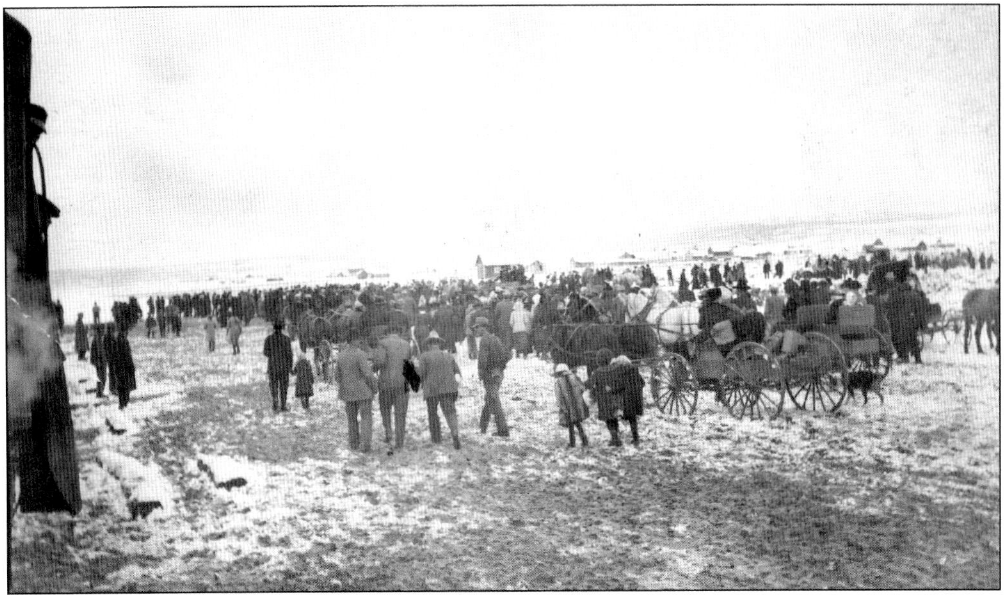

Railroad Day. After years of waiting, the first passenger train rolled into town. The pervasive excitement and camaraderie of that day is evident in this photograph. Most frontier towns on various railroad lines celebrated a Railroad Day. It was usually a momentous occasion because the arrival of the rails promised great changes to remote locales.

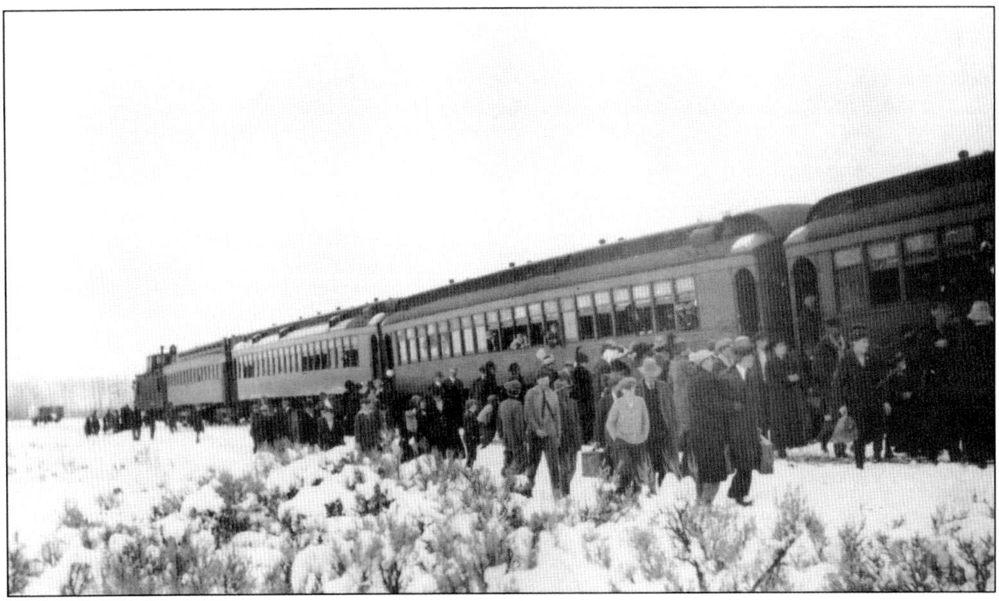

First Passenger Train in Craig, November 1913. Crowds gathered on a snowy morning, November 21, 1913, to see the first passenger train arrive. The event was a major happening for the region as it signaled improved travel, access to the Denver area, and increased ability to ship and receive goods, including mail.

Louise Wick, c. 1916. With the coming of the Moffat Road into Craig in 1913, local residents were able to travel easily to Denver and points beyond. Craig resident Louise Wick poses on the Craig train platform in front of the boxcar, which served as the station's depot until a brick depot was completed in 1917.

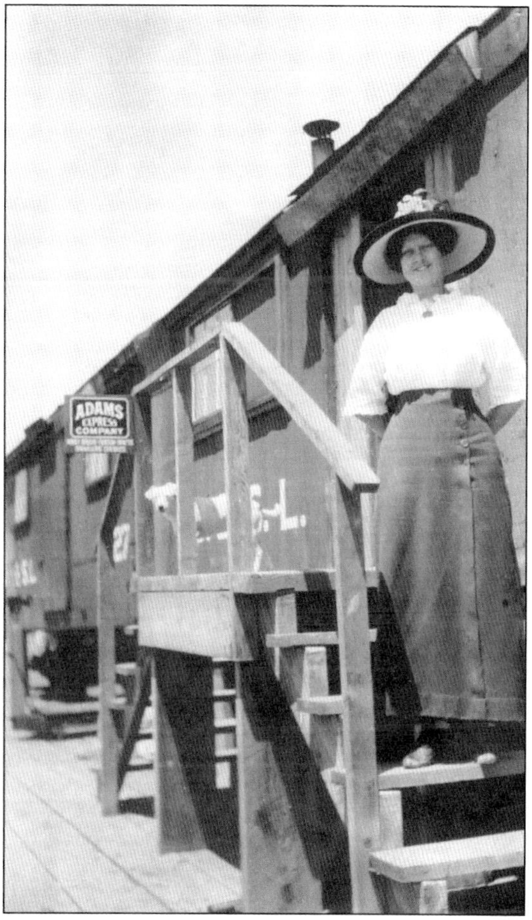

Train Derailment, 1914. This train derailment occurred east of Craig. The passenger train had departed Sunday, July 12, 1914, but derailed east of town. The rail line had been laid in heavy rains the previous fall and had either not settled or packed well at the time. This caused the ensuing derailment the following summer. (Photograph by Robert M. Richardson IV.)

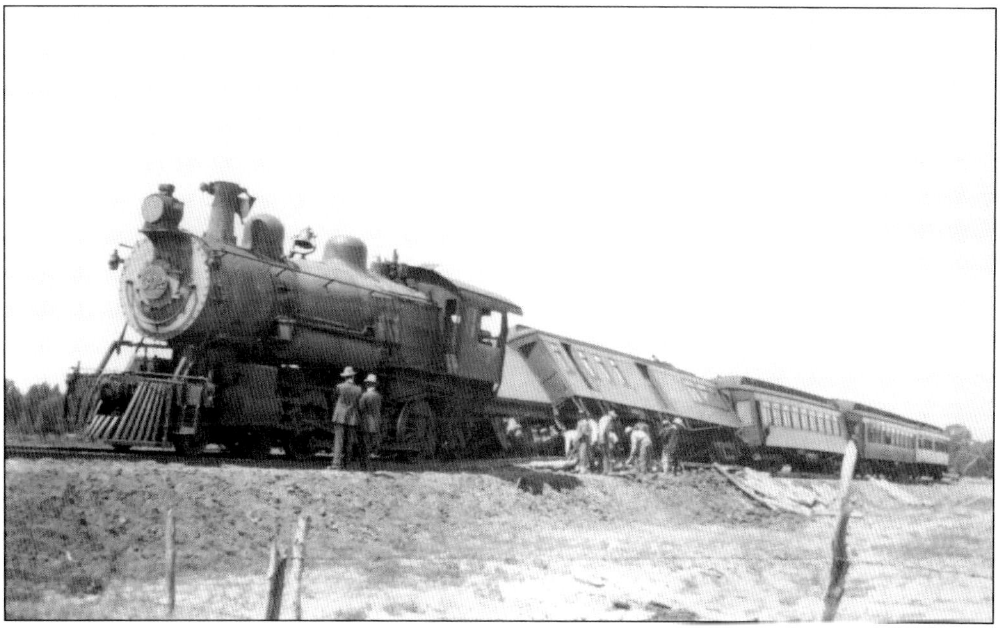

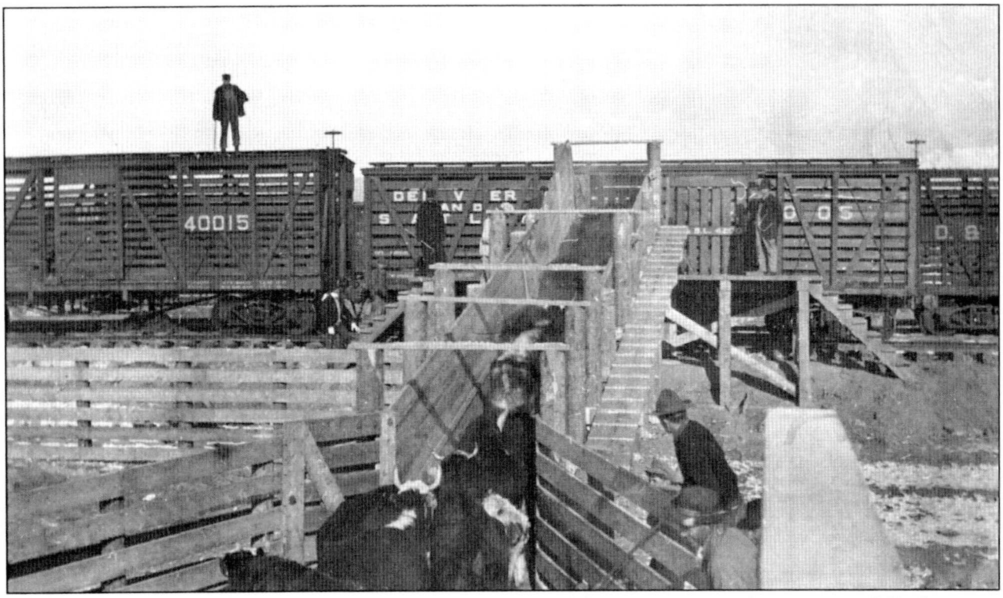

SHIPPING CATTLE, 1914. With the arrival of the new rail line into Craig the previous fall, shipping cattle—and other local goods—suddenly became much easier. In this photograph, men move the cattle up into the railcars at the Craig Stockyards.

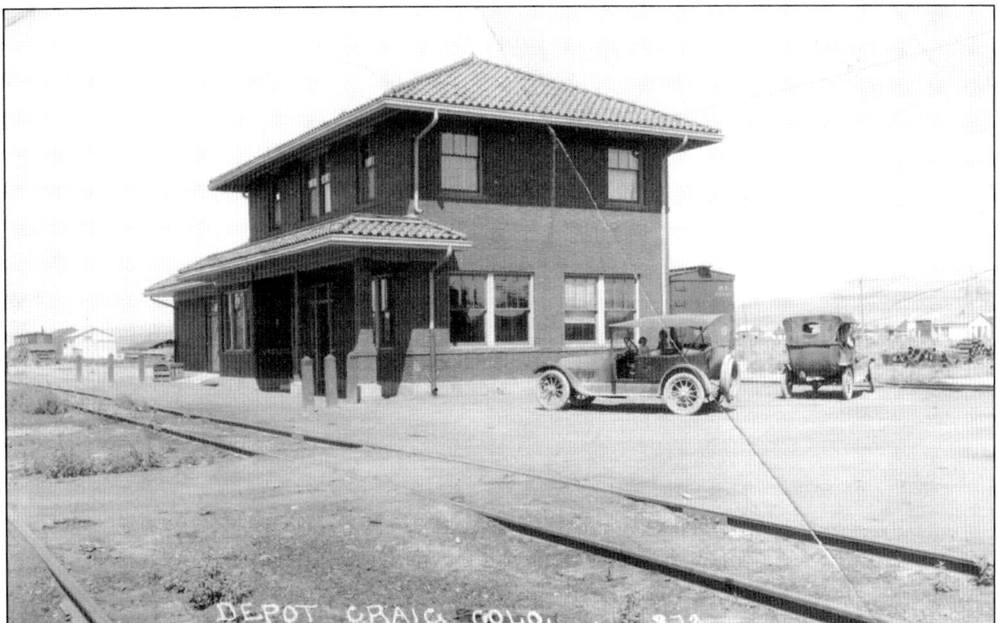

CRAIG RAILWAY STATION. The brick station, almost identical in layout to the Hayden station, was completed in the spring of 1917, four years after the arrival of the first passenger train in Craig. (Photograph by George Welch.)

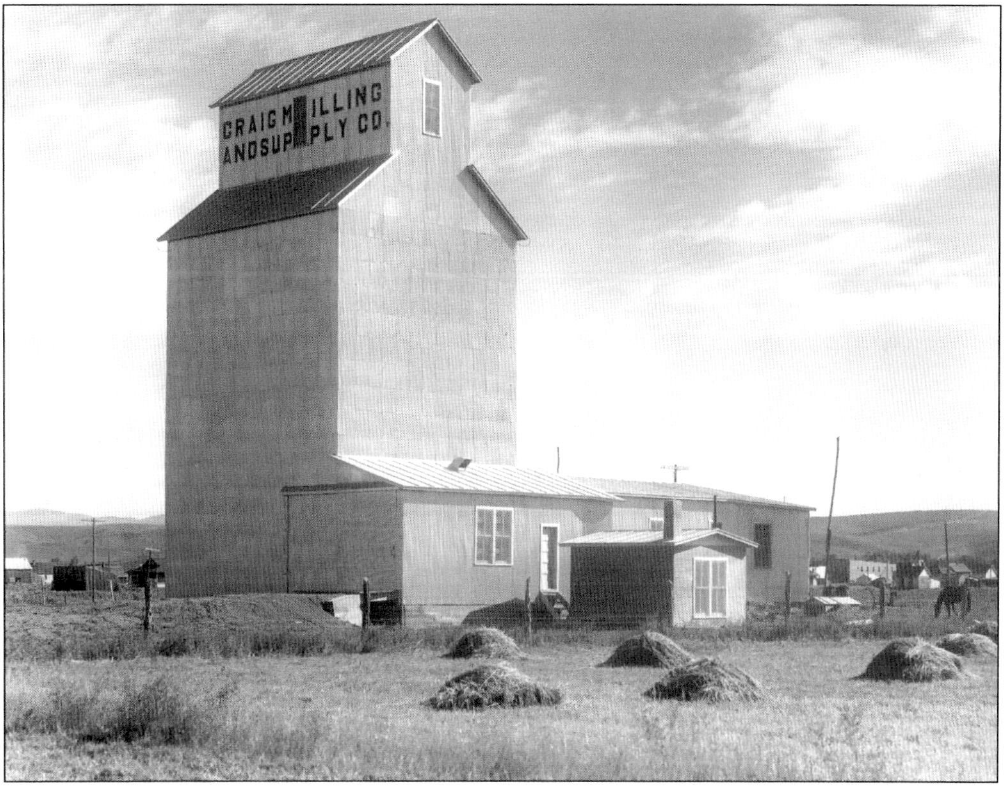

Craig Flour Mill, 1916. Photographer L.C. McClure traveled to Northwest Colorado in the summer of 1916. He took this image of the Craig Milling and Supply Company, which was located south of the railroad tracks on Ranney Street. This mill burned in November 1921. (Courtesy Denver Public Library.)

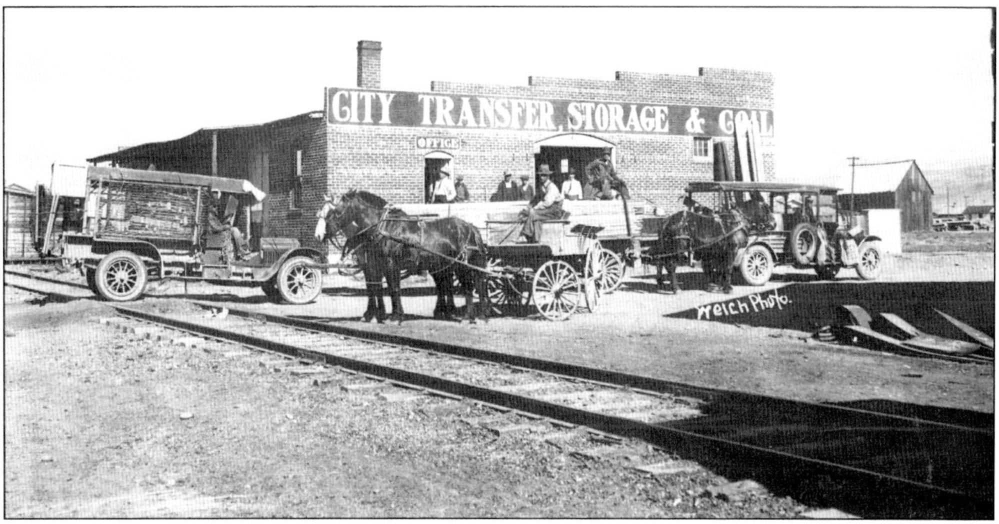

Changes in Transportation Methods, 1918. This is an assorted collection of transportation vehicles at the tracks of the Moffat Road. The railroad opened up the area to more goods from the outside. Here, vehicles of all types are moving everything from agricultural goods to lumber. (Photograph by George Welch.)

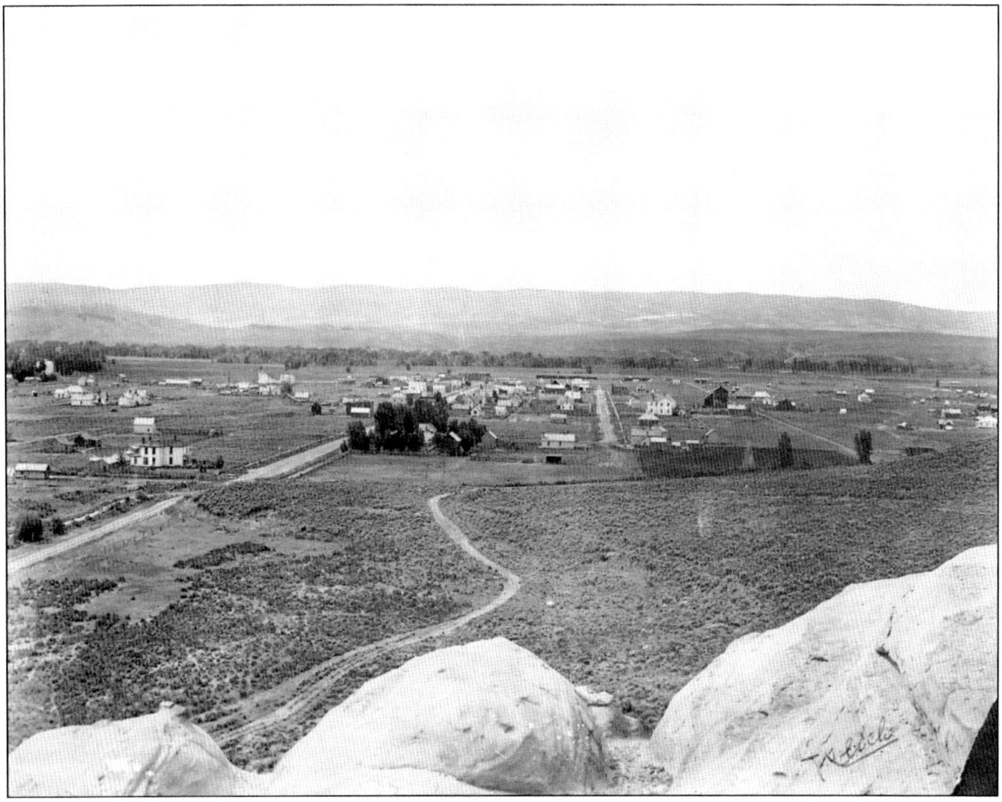

SUMMERTIME VIEW, 1914. Traveling photographer E.T. Davis of Out West Photos in Denver took this view of Craig looking south from the Sandrocks. He passed through the Yampa Valley in 1914.

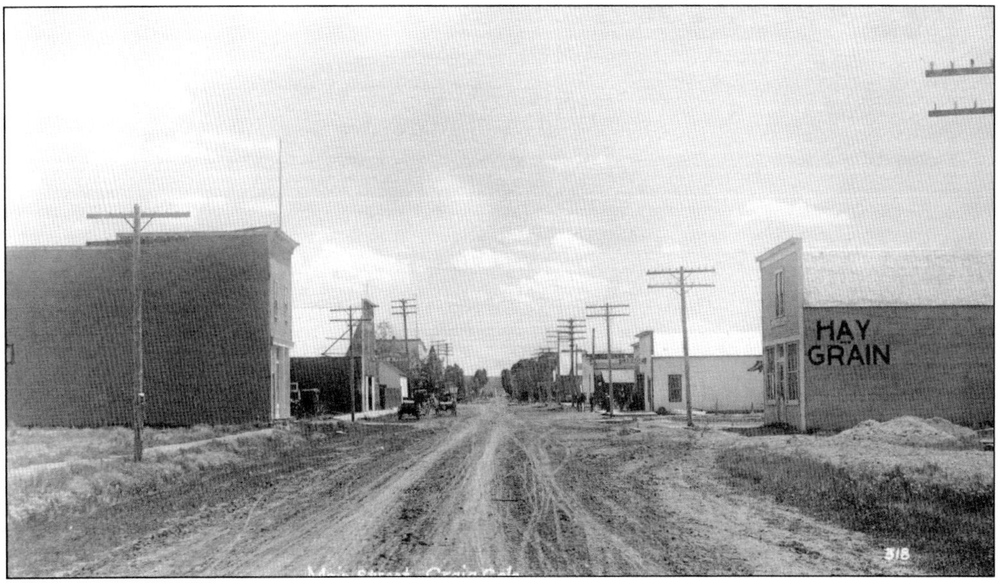

CRAIG, SUMMER 1914. Craig's Main Street (present-day Victory Way) is pictured looking to the east. (Photograph by E.T. Davis.)

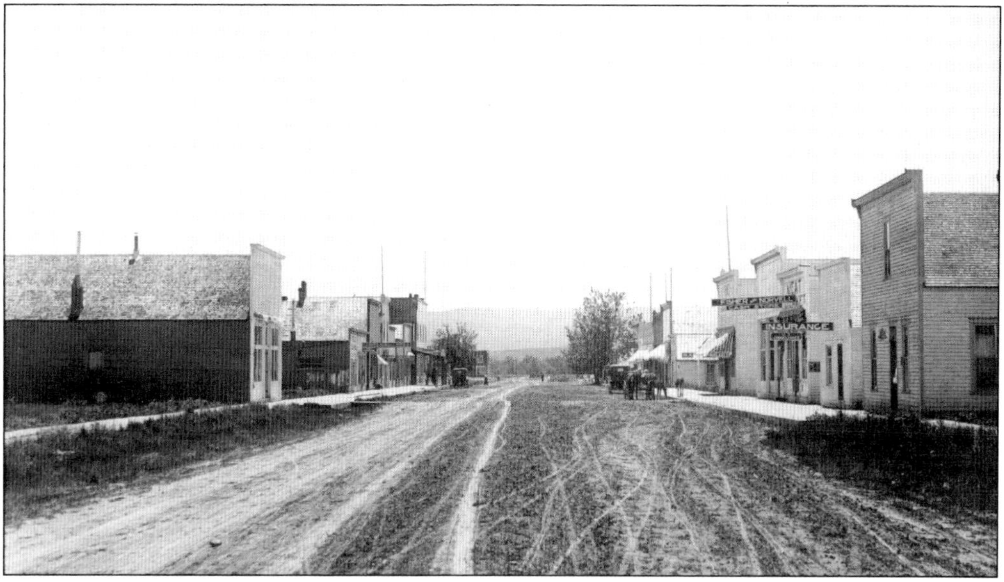

DOWNTOWN CRAIG, 1914. In addition to documenting the new concrete sidewalks, the view, looking south, shows the Fisher & Norvell Cash Store and the Bell Telephone building, both on the right. (Photograph by E.T. Davis.)

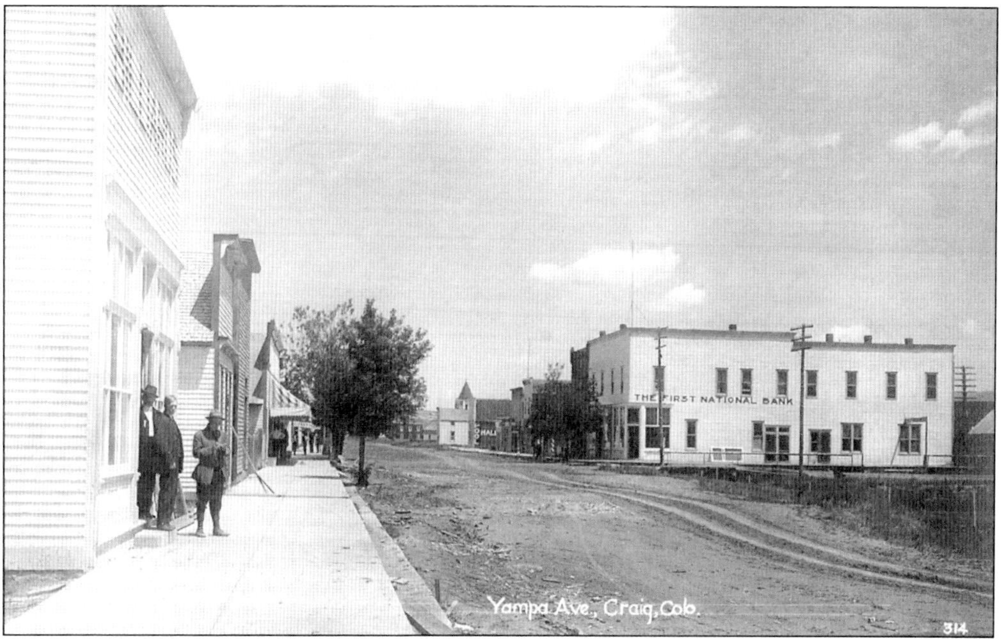

CRAIG, 1914. This photograph shows the First National Bank building on the right at the intersection of Yampa Avenue and Main Street. The bank moved out of this building in 1916 when a new brick building was finished across the street. (Photograph by E.T. Davis.)

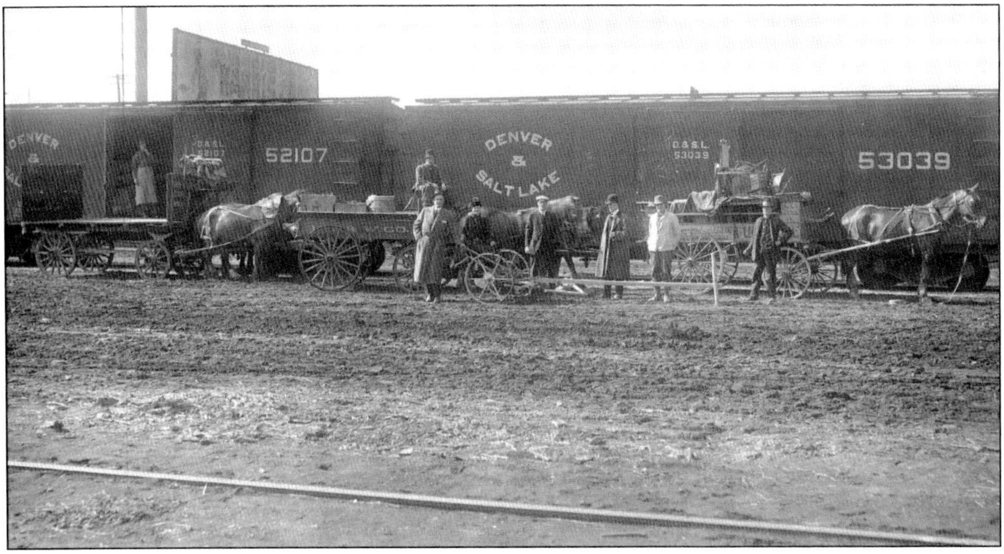

Hopeful Future, c. 1917. *Denver Post* editor Volney Hoggett (center left, foreground) vigorously promoted the Great Divide Townsite northwest of Craig. He stands with hopeful homesteaders at the Moffat Road station in Denver. The woman seated behind Hoggett wears an enthusiastic smile as she prepares to board the Denver & Salt Lake Train for Craig. Her household goods are piled high on the wagon at right. (Photograph by L.C. McClure; courtesy Wyman Living History Museum.)

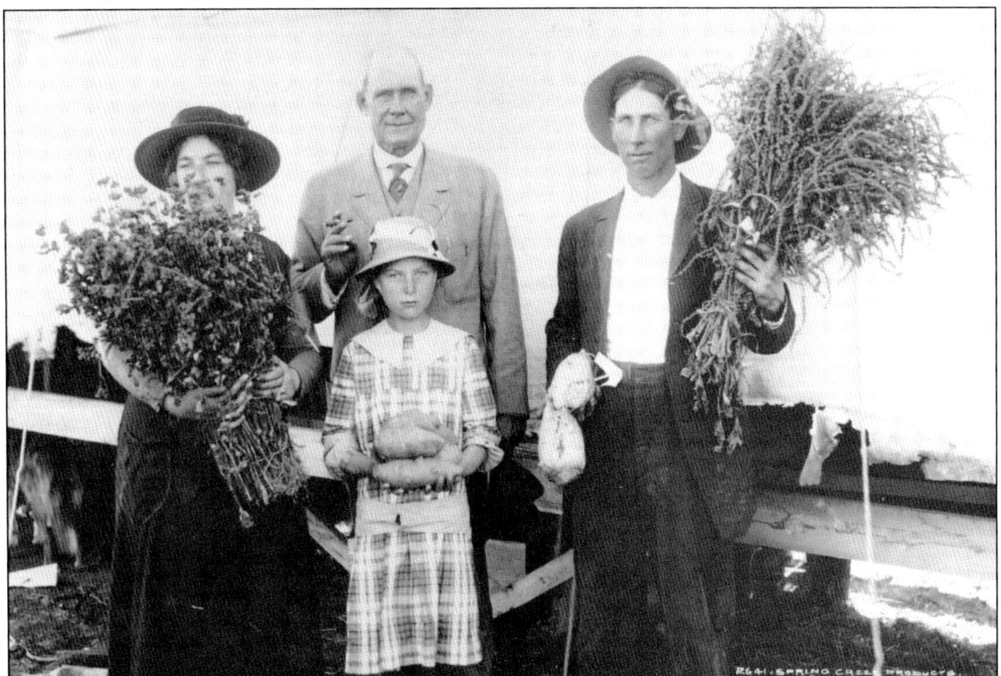

Brief Success, 1917. The Great Divide Townsite boomed for a few years when there was abundant moisture. When the weather settled back into a dry pattern, just prior to the Great Depression, many settlers were unable to prove-up on their land and abandoned their homestead efforts. Townsite promoter Volney Hoggett is pictured with homesteaders and their produce at a promotional fair at the townsite. (Photograph by L.C. McClure; courtesy Denver Public Library.)

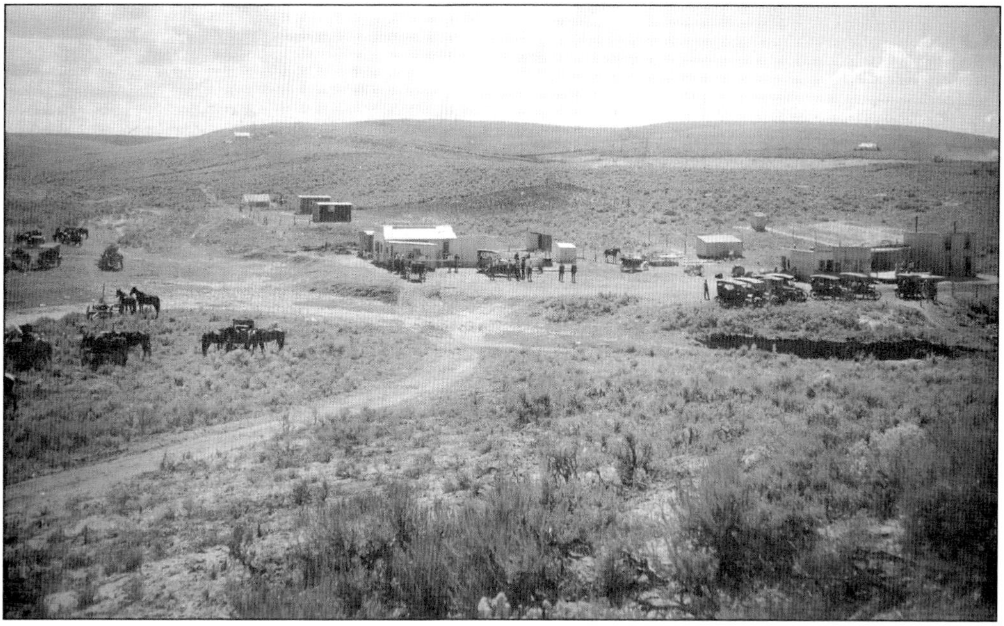

Great Divide Townsite, 1917. A partially completed hotel, post office, mercantile, and stacks of new lumber all proclaim a townsite that is going to grow. Even the community dump, on the right, is off to a good start. Despite hopes that this would become an agricultural center, the small community withered with the return of normal, dry years. This little town was later moved to higher ground. (Photograph by L.C. McClure; courtesy Wyman Living History Museum.)

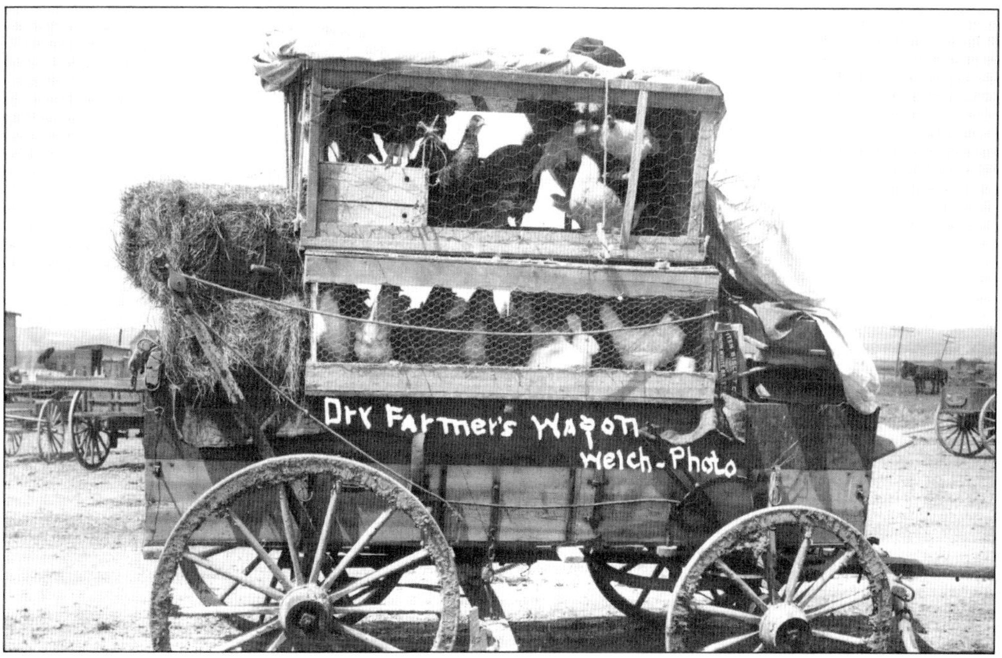

Dry Farmer's Wagon. This photograph was probably taken near the railroad tracks during the homestead boom of 1916 and speaks volumes about the conditions of the time. Immigrants at this time came into the area on railcars loaded with all the possessions and livestock they could afford to transport. (Photograph by George Welch.)

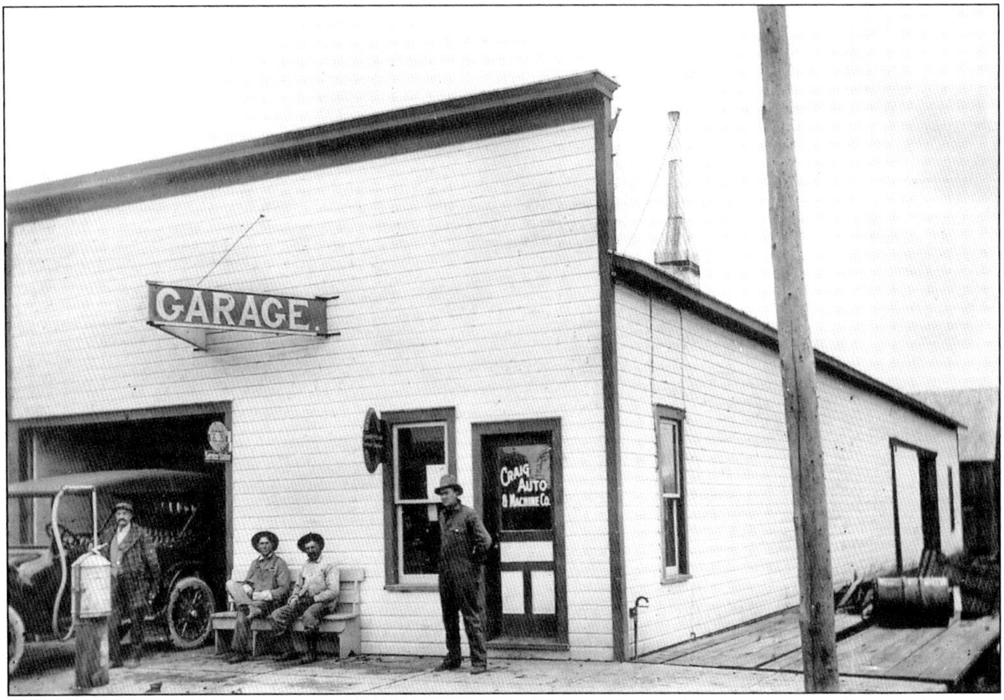

Early Craig Gas Station, c. 1915. Men pose outside the Craig Auto Garage on Yampa Avenue. Built in 1912, this was the first structure built in Craig expressly for the purpose of servicing automobiles. A very early gas pump is visible on the left. Within just two or three years, there was a rapid change from horse-drawn conveyances to automobiles throughout the region.

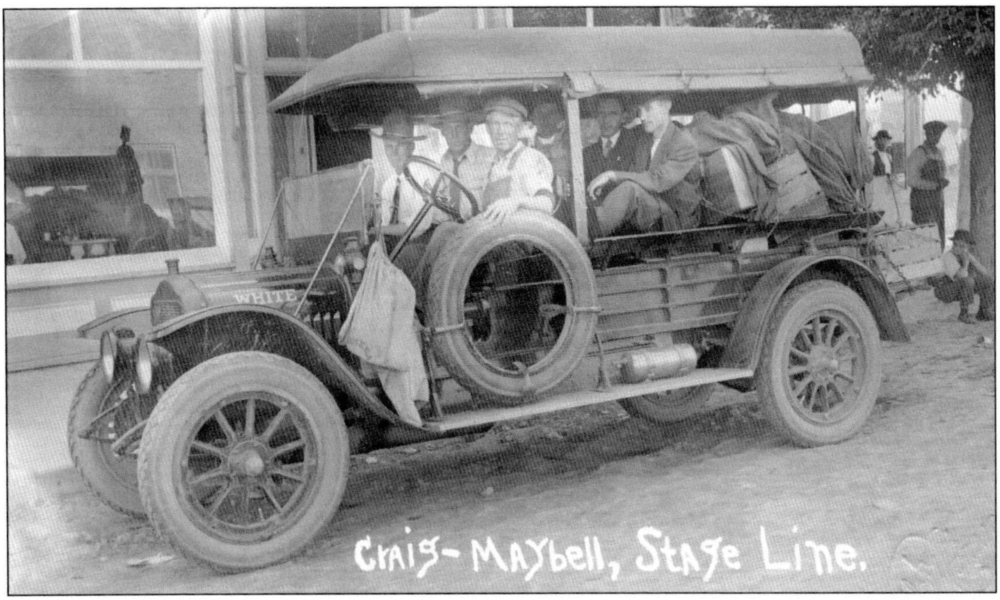

Maybell Stage, Summer of 1917. Tom Trevenen sits at the wheel of his White truck as he prepares to leave for Maybell with a load of passengers. With the coming of the automobile, the horse drawn stage was replaced with the auto stage. The stage also carried the US mail, as seen by the bag hanging at the front of the vehicle near the door. (Photograph by George Welch.)

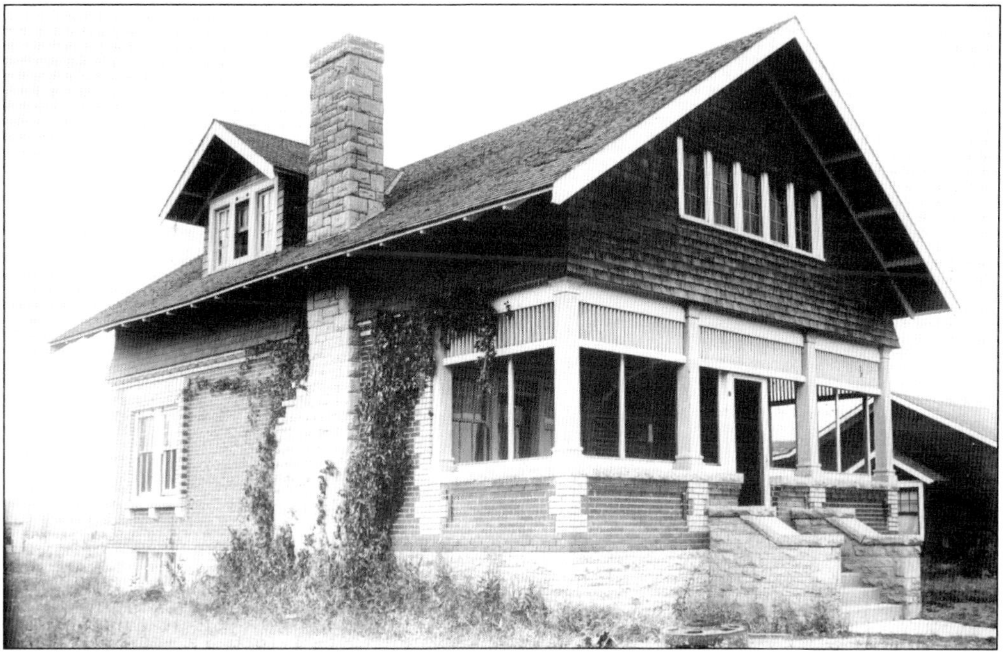

THE BAKER BUNGELOW. This lovely bungalow-style home on Breeze Street was built for Cora Baker and sometimes served as an overflow for her Baker Hotel on Yampa Avenue. It was constructed in 1917 for $11,000. (Photograph by George Welch.)

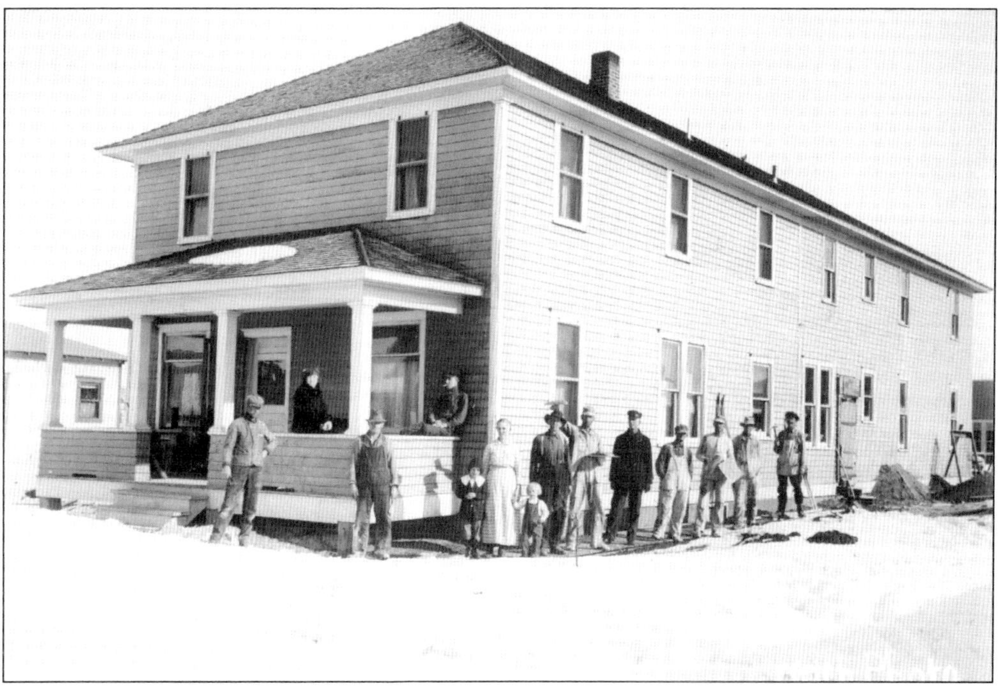

ADDITION TO THE DAVIS HOTEL, C. 1918. John and Molly Davis posed with their two children in front of their hotel at 537 School Street about 1918. The hotel, only a year or two old at the time, was already requiring an addition of 12 rooms. Several boarders and the construction crew are also pictured here. (Photograph by George Welch.)

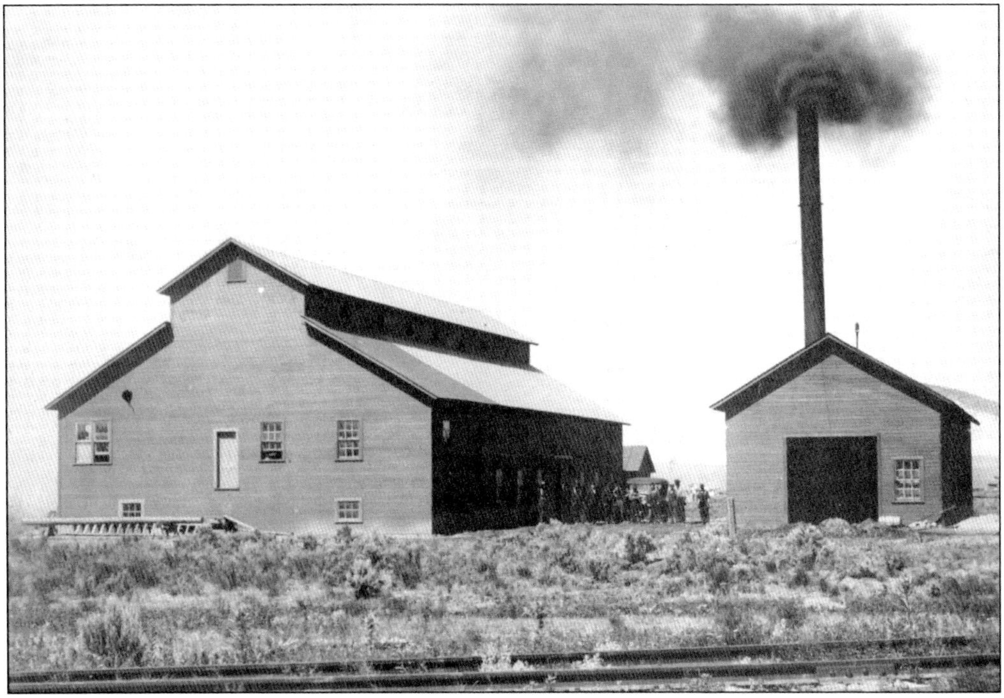

FRENCH PLOW WORKS, 1919. The French Plow factory was built in the 200 block of Russell Street south of the railroad tracks in 1918. With the influx of homesteaders in the 1910s, this manufacturing enterprise seemed like a sure bet. The factory produced plows for about two years before it closed due to financial problems. (Photograph by George Welch.)

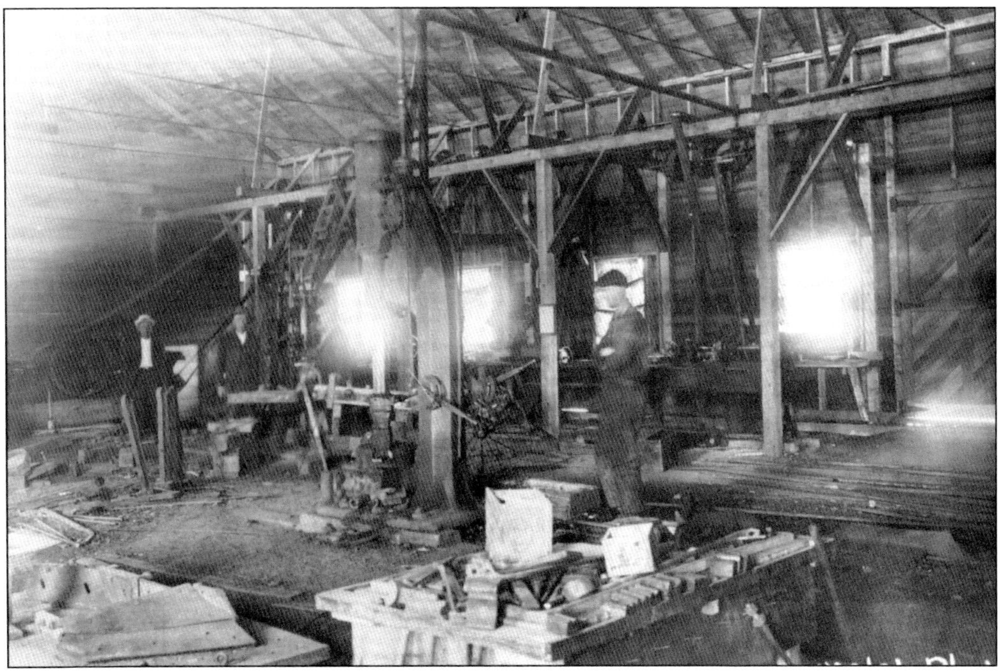

INTERIOR OF THE FRENCH PLOW FACTORY. This is the interior of the French Plow Factory, which operated between 1918 and 1920. (Photograph by George Welch.)

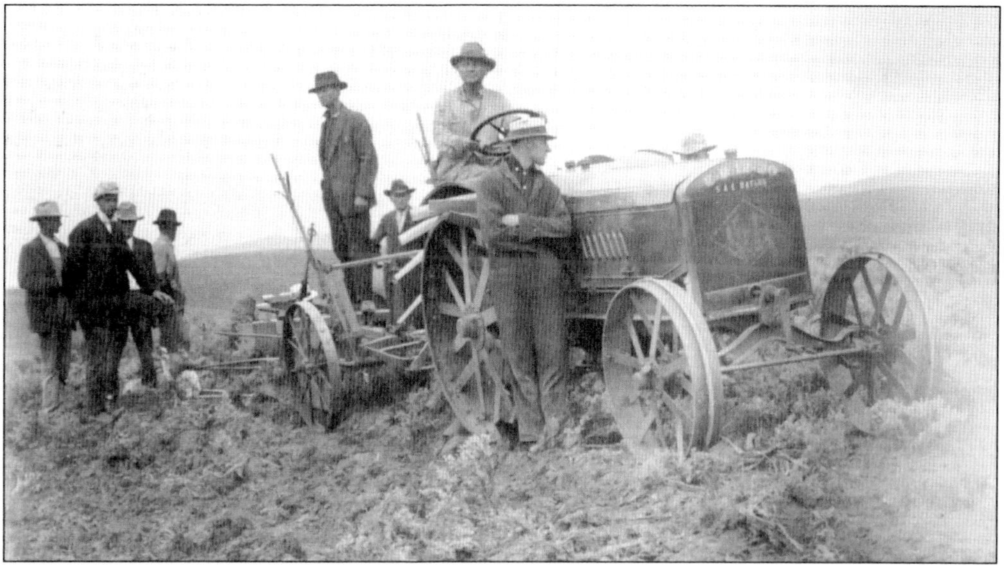

French Plow, 1919. This image shows a French plow being demonstrated at the Charles Van Dorn place west of Craig. The plows were sold locally and were shipped out by railroad to larger markets. (Photograph by George Welch.)

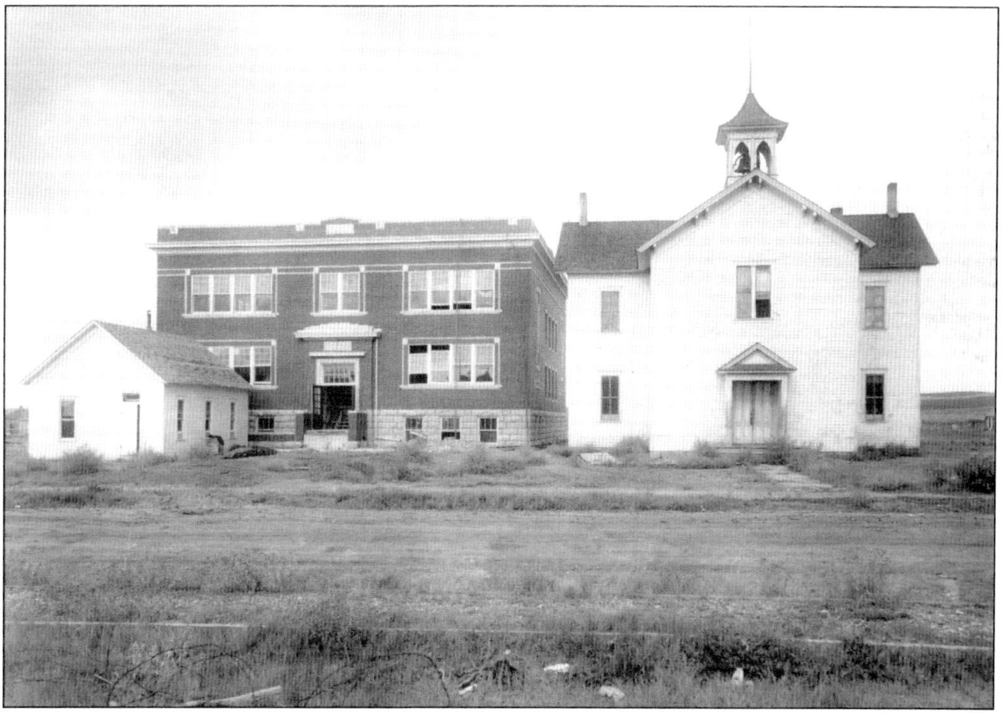

Three Craig Schools, 1916. This photograph documents three school buildings on Breeze Street. The large white building was erected in 1892 to replace the first log school. The small structure on the left was built to serve as an overflow building in the 1910s, when a new wave of homesteaders caused school enrollment to skyrocket. The new brick school was being completed in 1916; later called the Breeze School, it was used until it was demolished in 1981. (Photograph by L.C. McClure; courtesy Denver Public Library.)

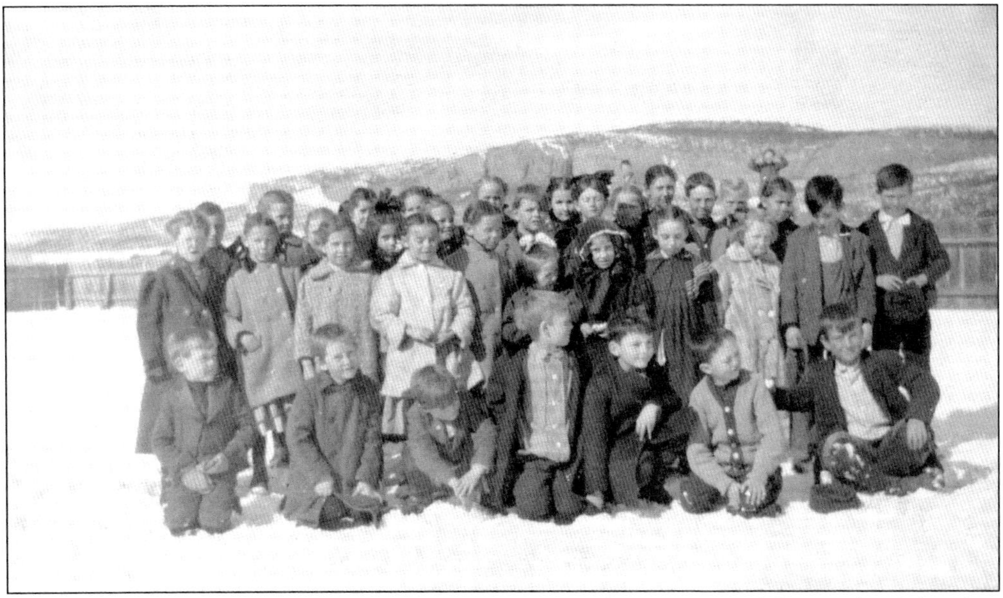

VALENTINE'S DAY, 1911. Clarence Welch, son of photographer George Welch, posed with his class outside the Craig School in the winter of 1911. The photograph was used on a Valentine's Day card.

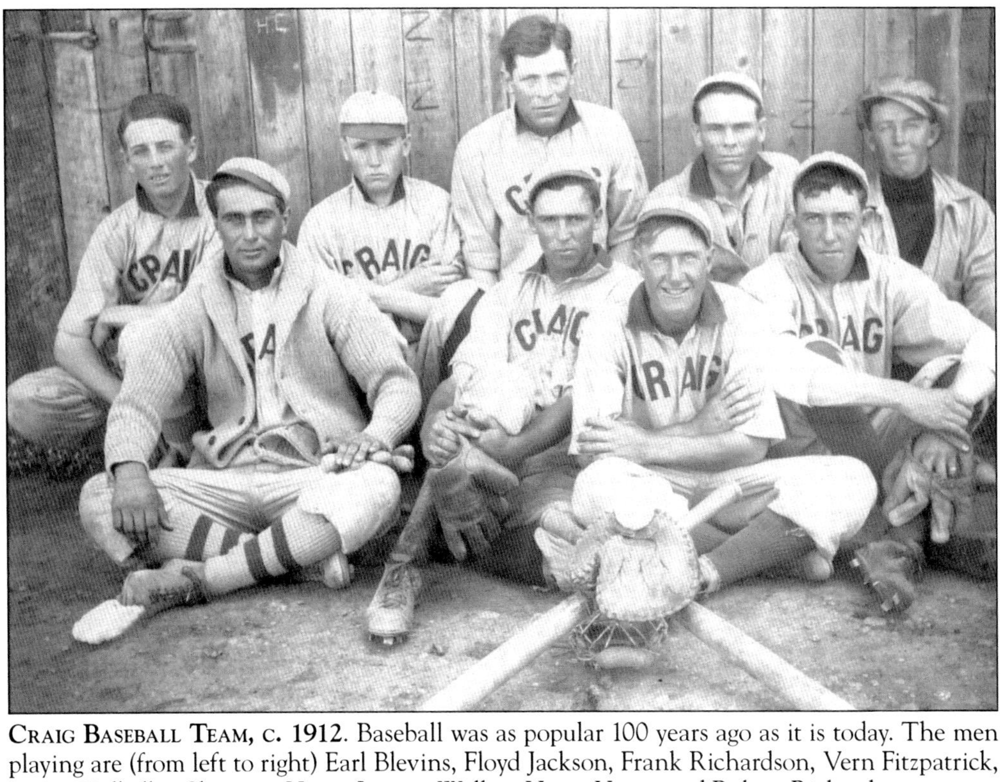

CRAIG BASEBALL TEAM, C. 1912. Baseball was as popular 100 years ago as it is today. The men playing are (from left to right) Earl Blevins, Floyd Jackson, Frank Richardson, Vern Fitzpatrick, Henry Kolbalba, Sherman Hyatt, Jimmie Walker, Harry Hyatt, and Robert Richardson.

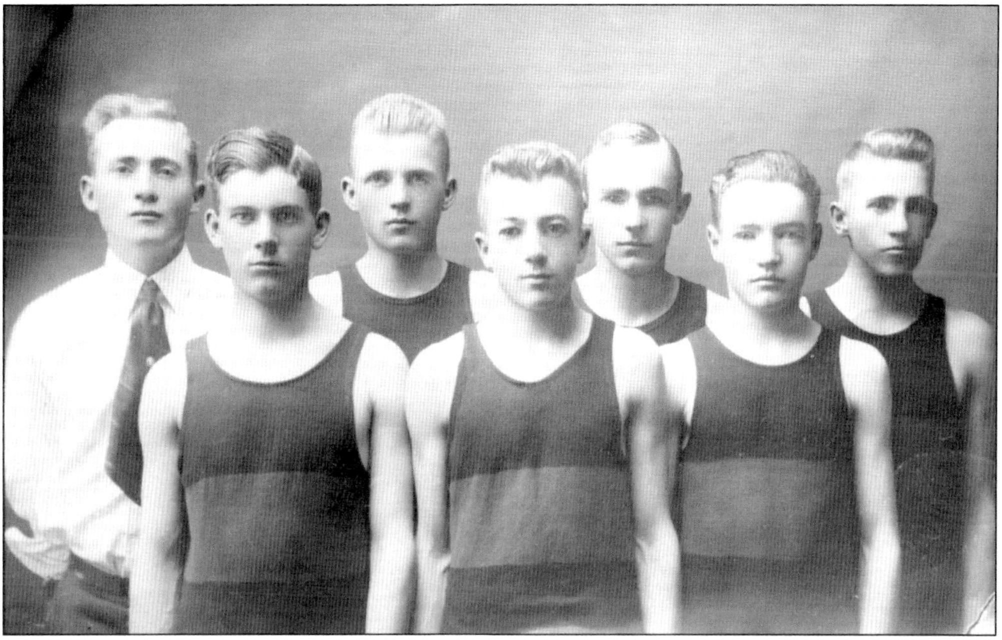

TRACK TEAM, C. 1911. Photographer George Welch took numerous photographs of local school sports teams that traveled the valley for competitions. Here, the Craig High School track team members are, from left to right, (first row) Clarence McKay, Frank White and Milton Taylor; (second row) Bill White, Charles White, Percy Hamilton, and ? Brown.

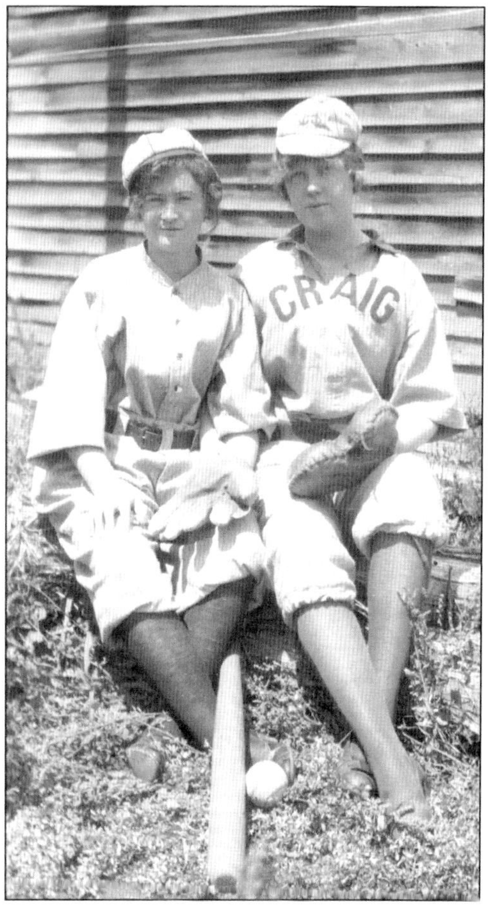

READY TO PLAY, C. 1911. Craig High School students and best friends Pearl McKay and Ellen Wick pose in uniforms from the Craig men's baseball team.

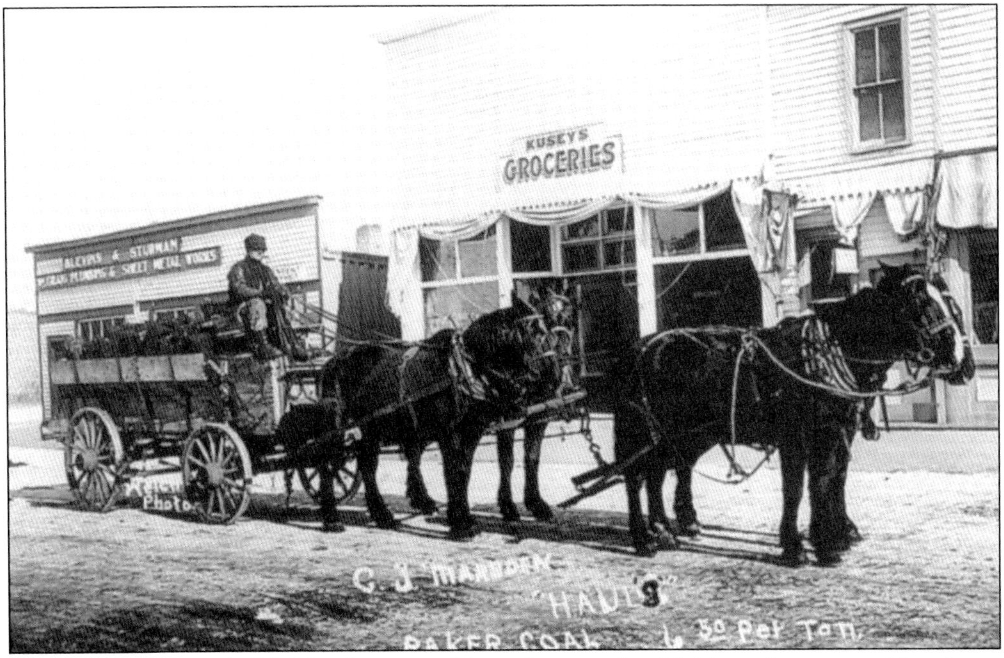

Hauling Coal, 1916. C.J. Marsden paused for a photograph with his coal wagon and four-horse team on Yampa Avenue. Blevins & Sturman Plumbing & Sheet Metal Works and Kusey's Groceries are both visible behind the wagon and team. (Photograph by George Welch.)

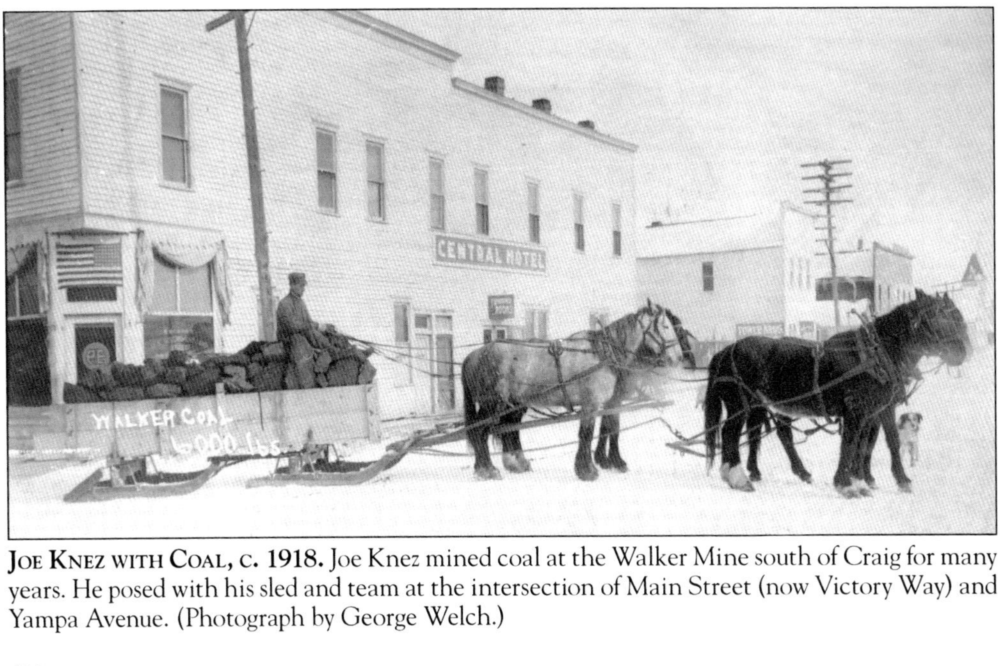

Joe Knez with Coal, c. 1918. Joe Knez mined coal at the Walker Mine south of Craig for many years. He posed with his sled and team at the intersection of Main Street (now Victory Way) and Yampa Avenue. (Photograph by George Welch.)

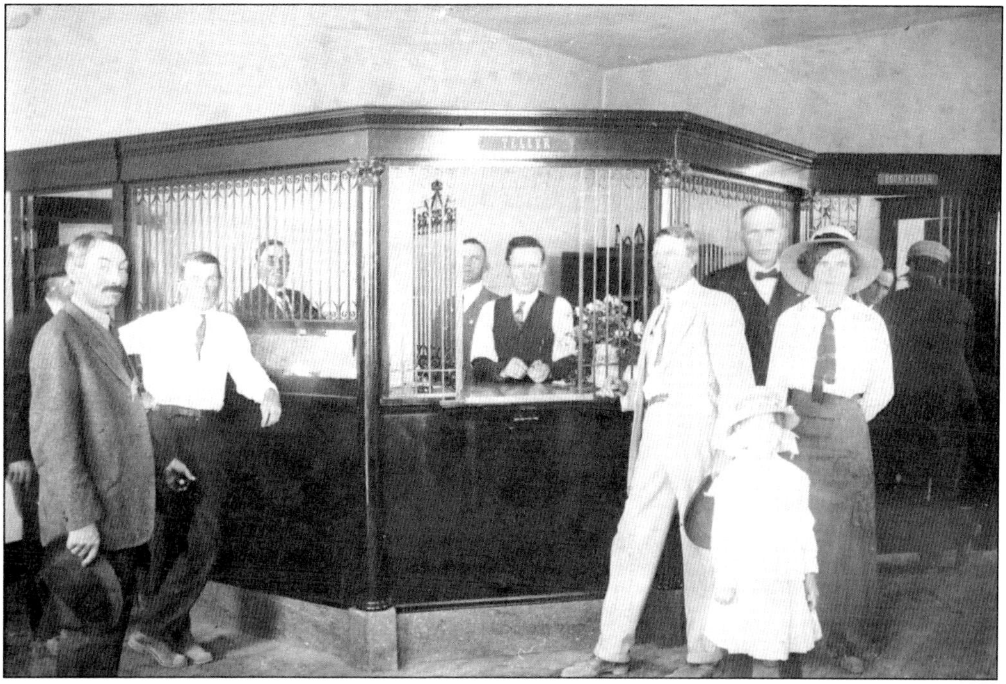

Banking Business. This bank changed names several times, starting out as the Bank of Amsterdam before becoming the Citizens Bank. In 1914, it became the First National Bank, with C.A. Van Dorn, president, shown here behind the teller's cage along with, from left to right, Lee Jones, Guy Ambrose, and Earl Van Tassel.

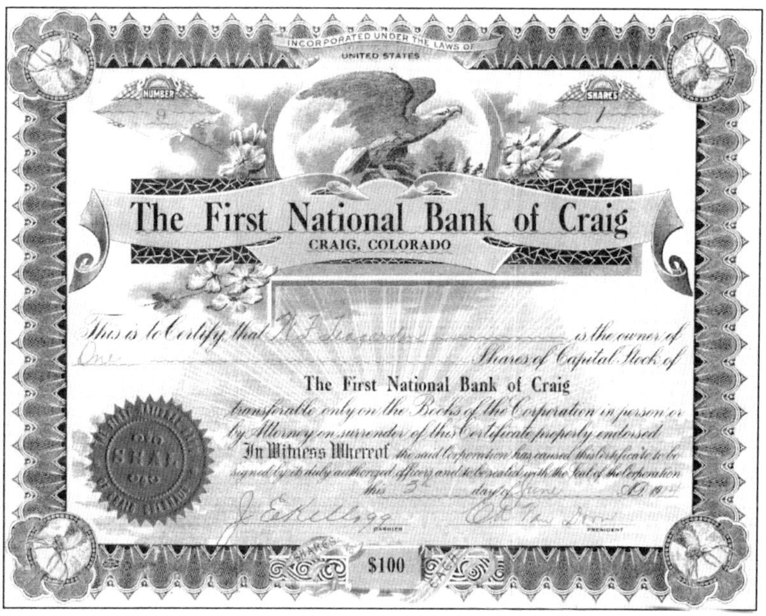

INCORPORATED UNDER THE LAWS OF UNITED STATES
NUMBER 9
SHARES 1
The First National Bank of Craig
CRAIG, COLORADO
This is to Certify that W.F. Teagarden is the owner of One Shares of Capital Stock of
The First National Bank of Craig
transferable only on the Books of the Corporation in person or by Attorney on surrender of this Certificate properly endorsed.
In Witness Whereof the said Corporation has caused this Certificate to be signed by its duly authorized officers and to be sealed with the Seal of the Corporation
this 3rd day of June A.D. 1914
SEAL
CASHIER
PRESIDENT
SHARES $100

Bank Certificate. This certificate was issued to W.F. Teagarden in 1914 when the Citizens Bank was nationalized and became the First National Bank of Craig.

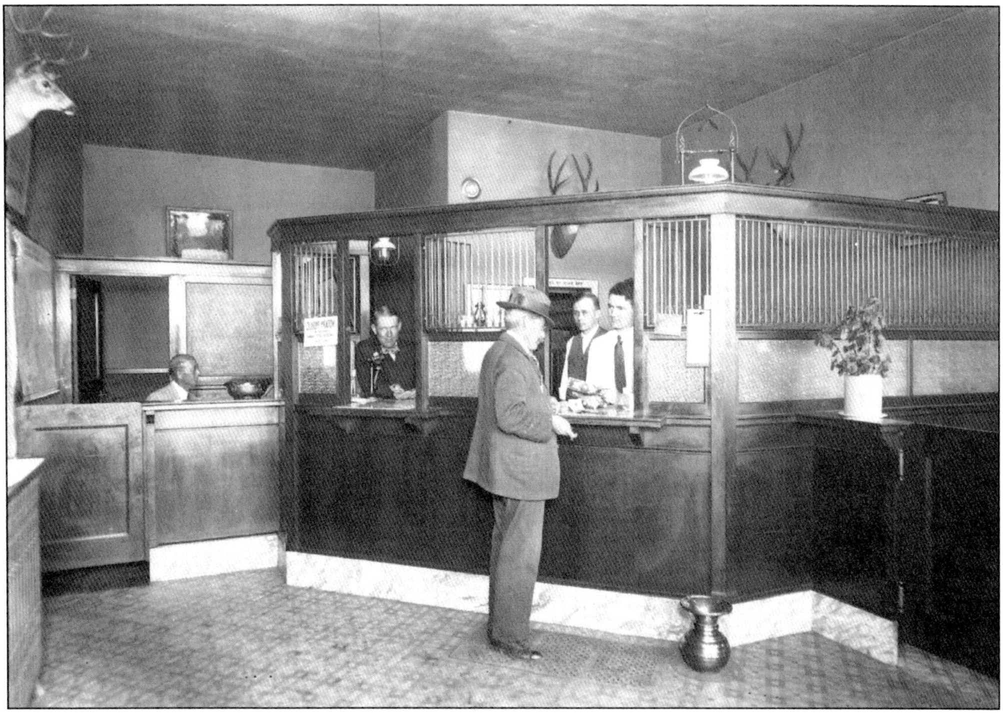

Interior of Craig National Bank, 1915. Local men posed for their photograph inside the bank, which was adjacent to the J.W. Hugus store at Yampa Avenue and Main Street. The bank, which had started operation in 1892, was the first one in Craig. It closed in 1932 during the Great Depression.

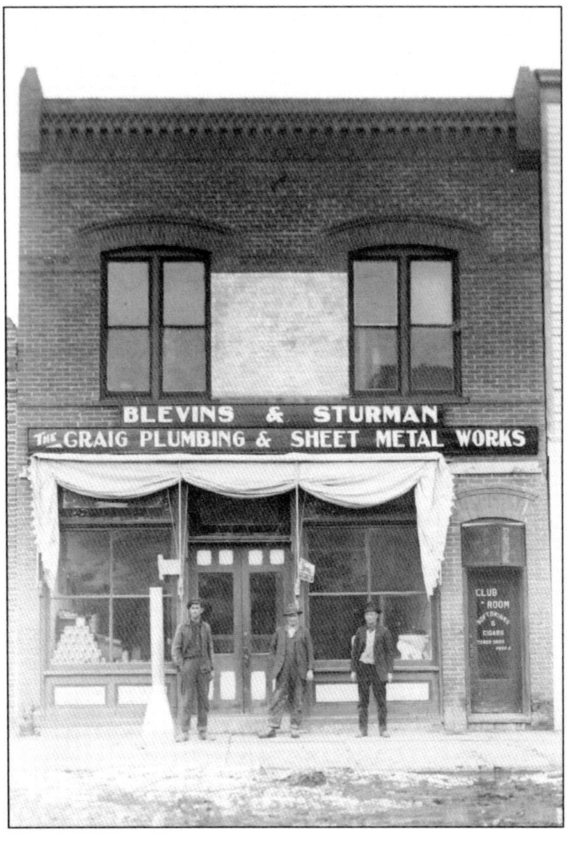

Craig Plumbing and Sheet Metal Works, 1920. Dick Sturman (left) posed with Jack Alfred and Earl Blevins in front of their business, Craig Plumbing and Sheet Metal Works, at 509 Yampa Avenue. The building, still in use today, was constructed in 1895 using locally produced bricks. George Welch used this photograph in a promotional booklet about Craig that was published in April 1920.

PEARL AIKEN (WHITE), 1910. The daughter of area homesteaders, Pearl moved to Craig in 1904 at age eight with her parents. She married Ralph White in 1912, and they raised three children here. They were heavily involved in many areas of the community throughout their lives.

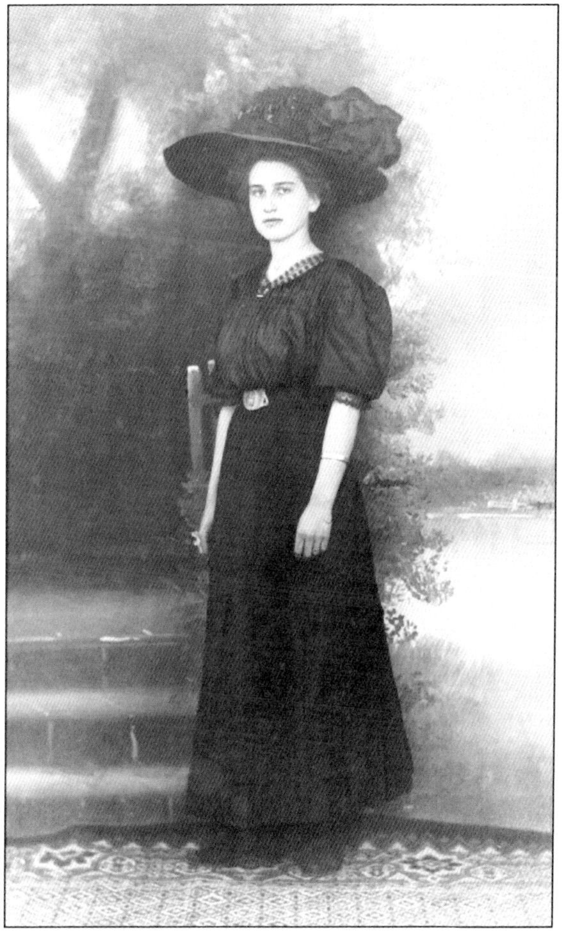

COMMUNITY CHRISTMAS TREE, 1919. The holiday atmosphere prevailed on Christmas Eve. Excited children gathered around the decorated tree at the intersection of Main Street and Yampa Avenue. This festivity was a tradition for a number of years in Craig. (Photograph by George Welch.)

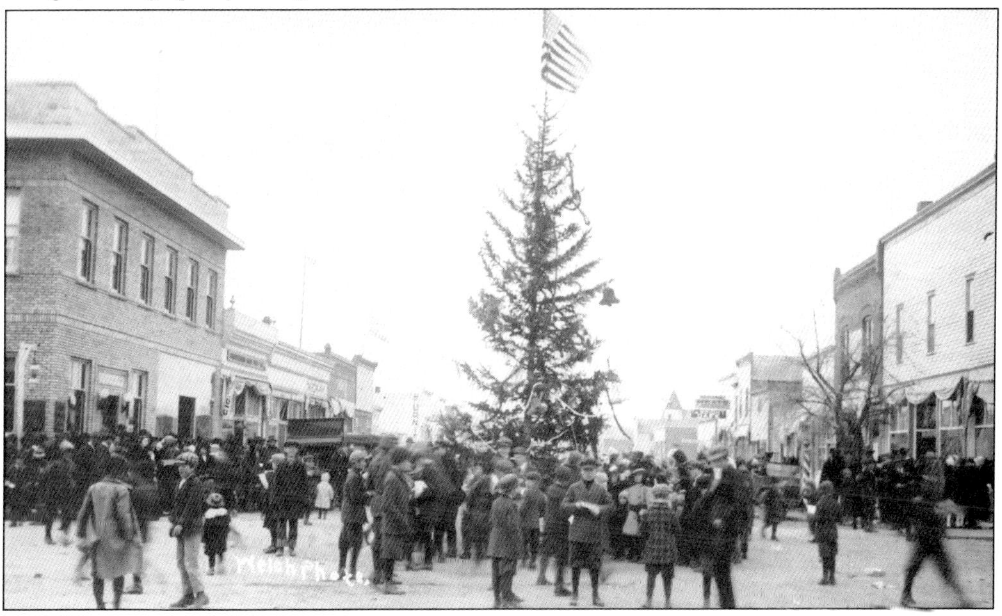

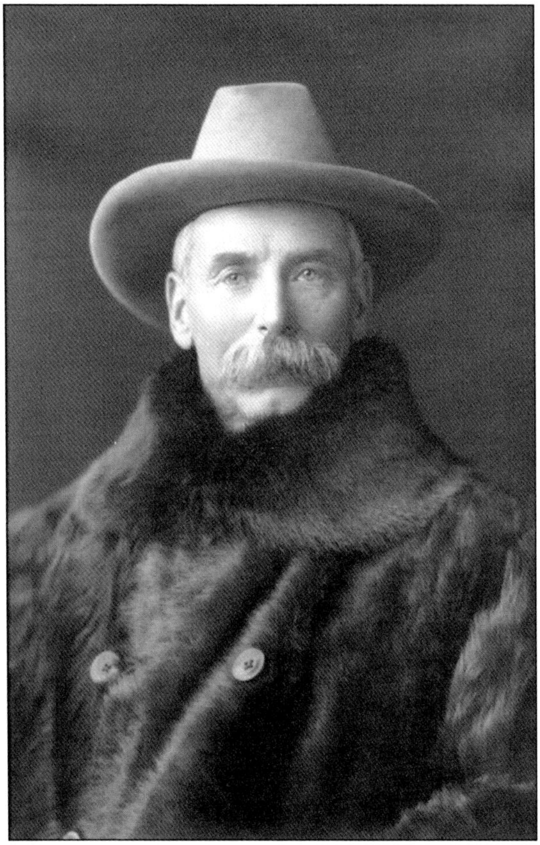

ARCHIE MCLACHLAN, C. 1908. McLachlan came to Craig from Prince Edward Island in Nova Scotia with Albert Ryan and brothers David and Donald Taylor in 1883. He homesteaded north of Craig and ran a sawmill on Black Mountain. McLachlan was elected in 1906 as a representative in the Colorado State House and served from 1907 to 1908. (Photograph by Kirkland of Denver.)

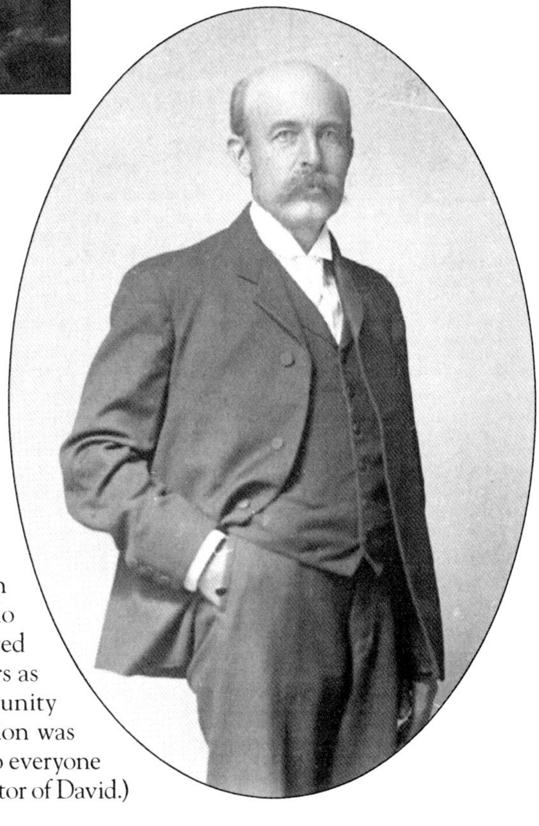

ROBERT V. BRYAN, MAYOR OF CRAIG. Born in 1854, Bryan came to northwestern Colorado in 1885, first settling near Maybell. He moved to Craig in 1891 and served in the early years as mayor. Bryan was so beloved by the community that on the day of his funeral, a proclamation was issued calling for all businesses to be closed so everyone could attend the services. (Photograph by Castor of David.)

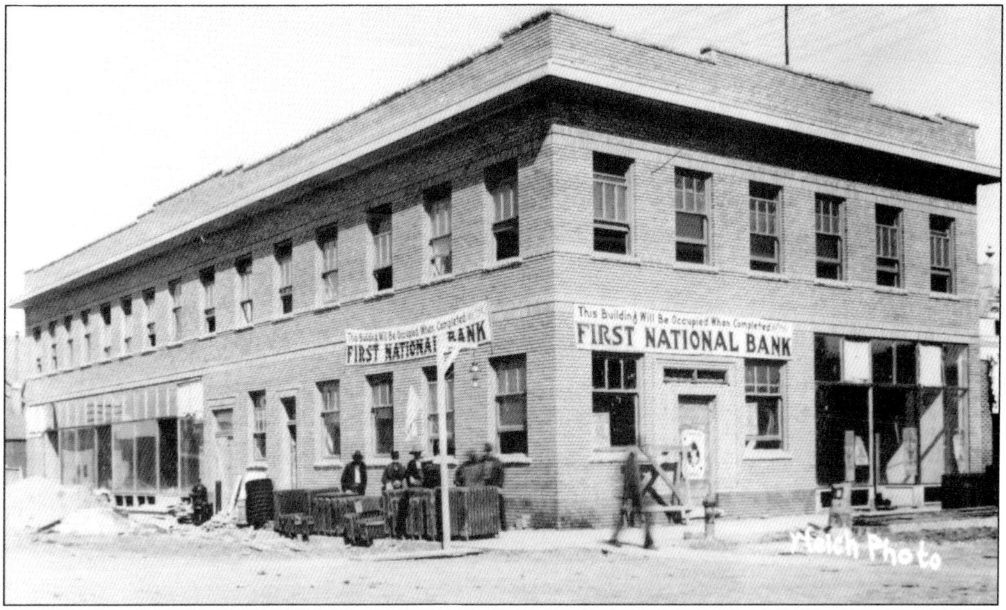

FIRST NATIONAL BANK BUILDING, 1916. The completed bank building, pictured here, is still located today at the northwest intersection of Yampa Avenue and Victory Way. During its 95 years in existence, this building has been used for many commercial enterprises. (Photograph by George Welch.)

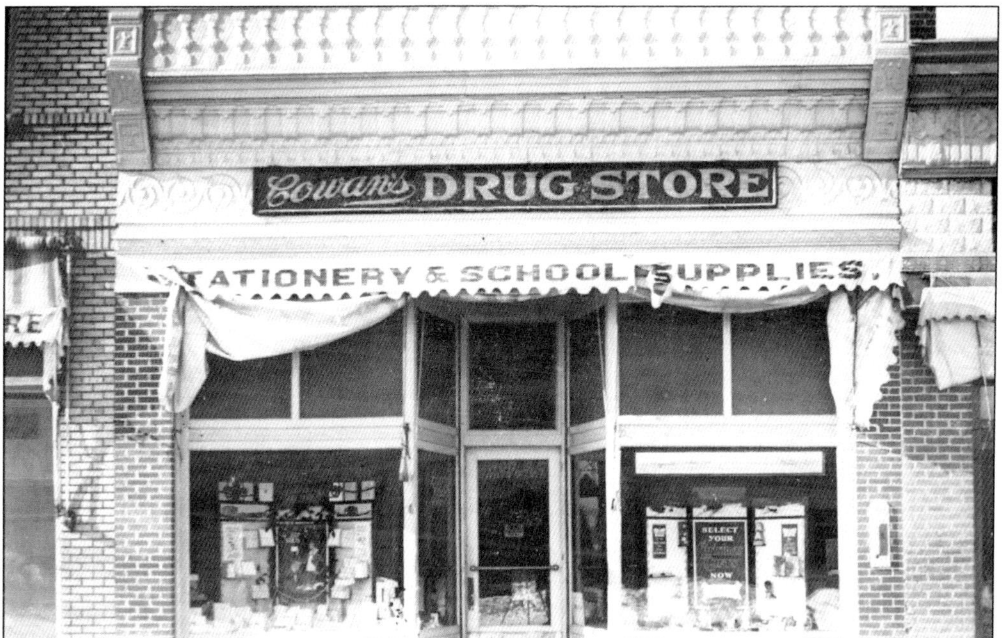

COWAN'S DRUG STORE, 1919. The drugstore is pictured shortly after Frank Cowan purchased it from the Chamberlain Gray drugstore chain. This business operated for more than 60 years in Craig. (Photograph by George Welch.)

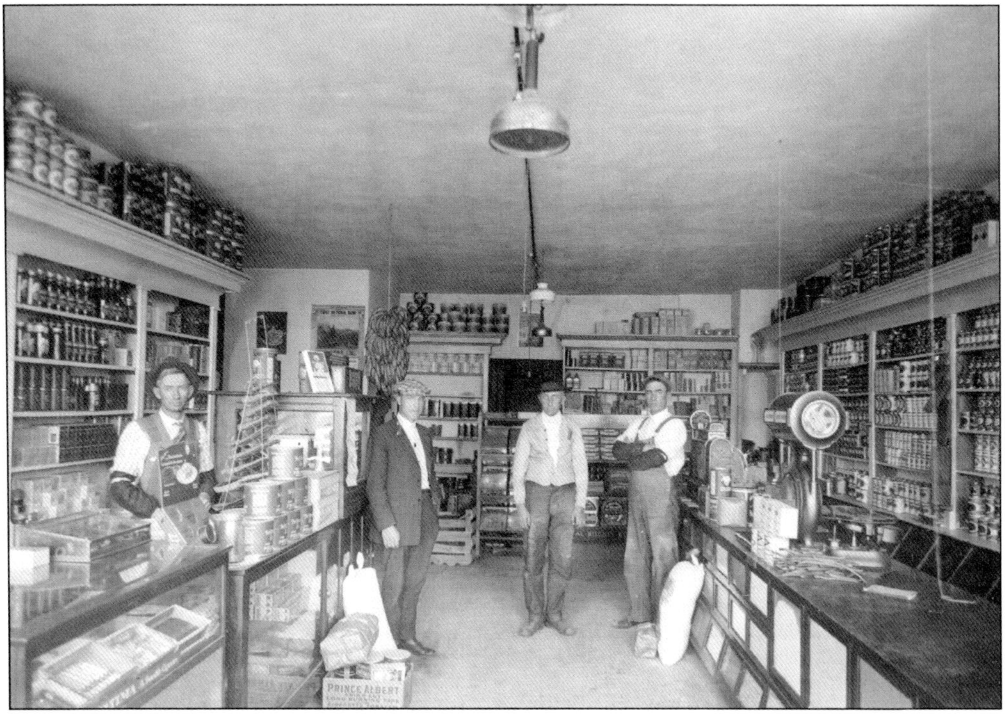

CRAIG STORE, 1920. This photograph shows Clyde Downs behind the counter of his general mercantile store, Haubrich and Downs. The store was located at 129 West Main Street (present-day West Victory Way).

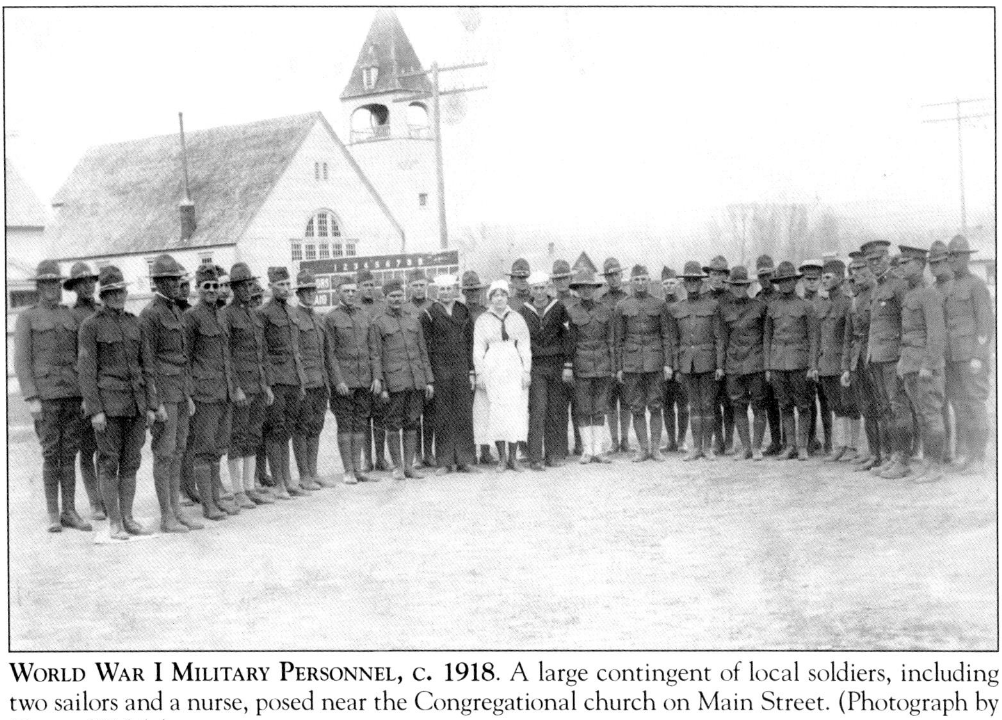

WORLD WAR I MILITARY PERSONNEL, C. 1918. A large contingent of local soldiers, including two sailors and a nurse, posed near the Congregational church on Main Street. (Photograph by George Welch.)

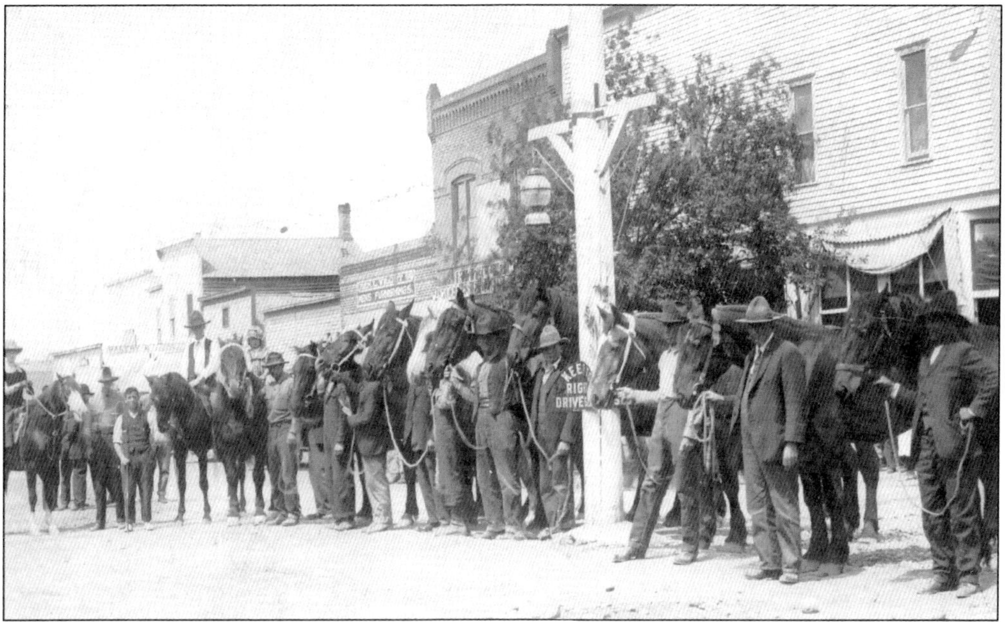

HORSE FAIR, MAY 1919. Men stand by Percheron breeding stock in front of the old First National Bank building at the northeast corner at Main Street and Yampa Avenue. The horses, from the Gossard Ranch at Axial, were brought to Craig for a horse fair held in May 1919. (Photograph by George Welch.)

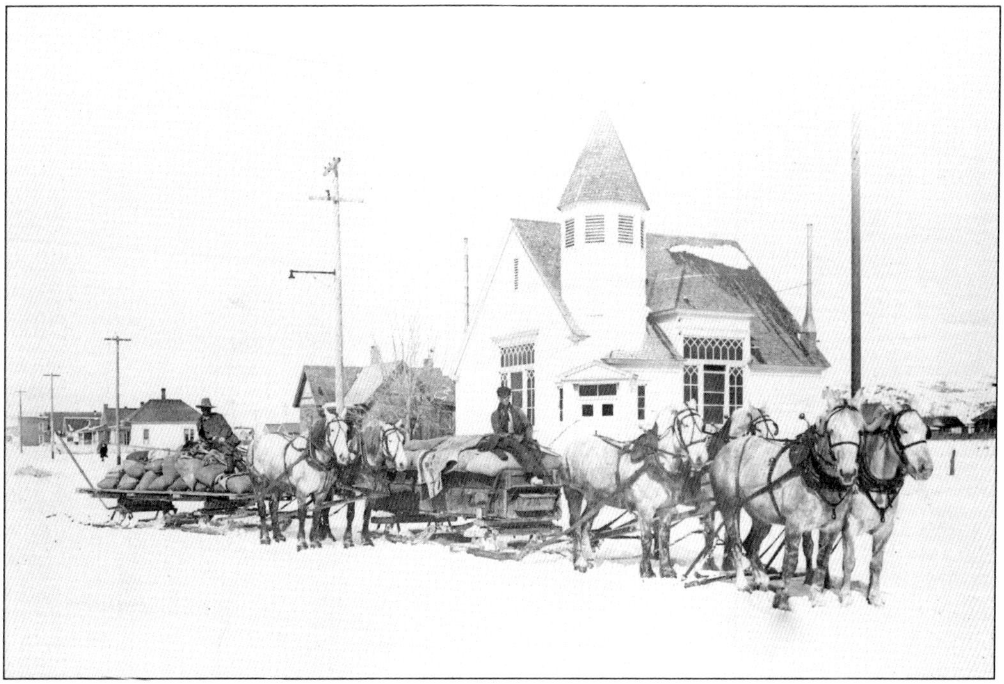

HAULING GRAIN, 1920–1921. Albert Horton (left) and Howard Lynch brought grain to Craig by sled during the winter of 1920–1921. The power lines were new to the Craig scene, having been installed in 1918. In the beginning, electrical service was only available during the day. (Photograph by George Welch.)

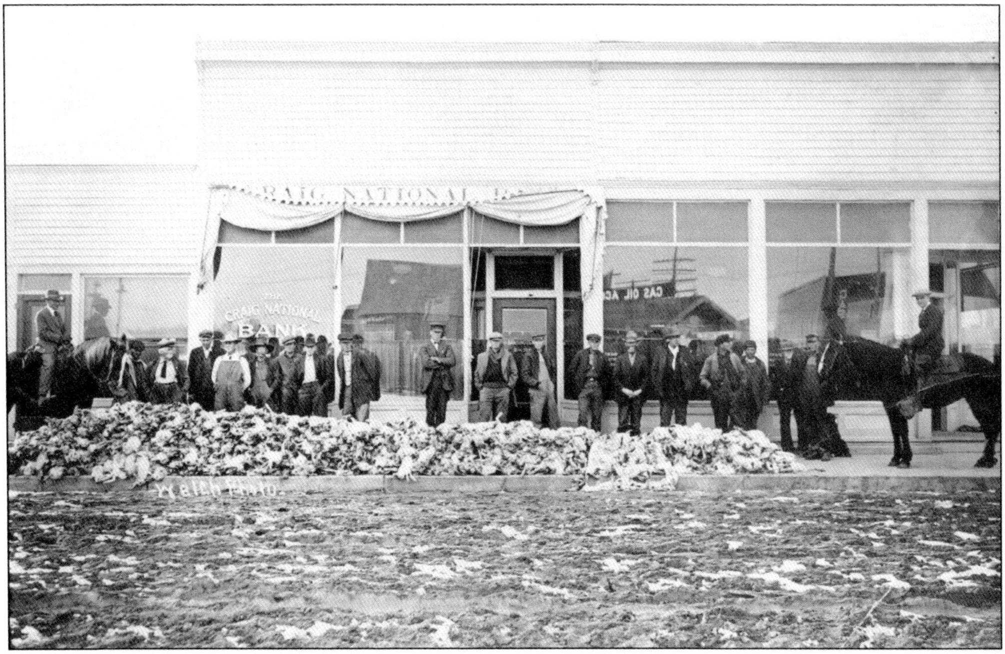

Rabbit Hunt. Rabbit hunts were frequently held in the early years to decrease their large population. The meat was not wasted; it was sent to Denver to feed the poor. (Health regulations would certainly not allow this today.) Pictured here is the Craig National Bank, which was located in the old J.W. Hugus building. (Photograph by George Welch.)

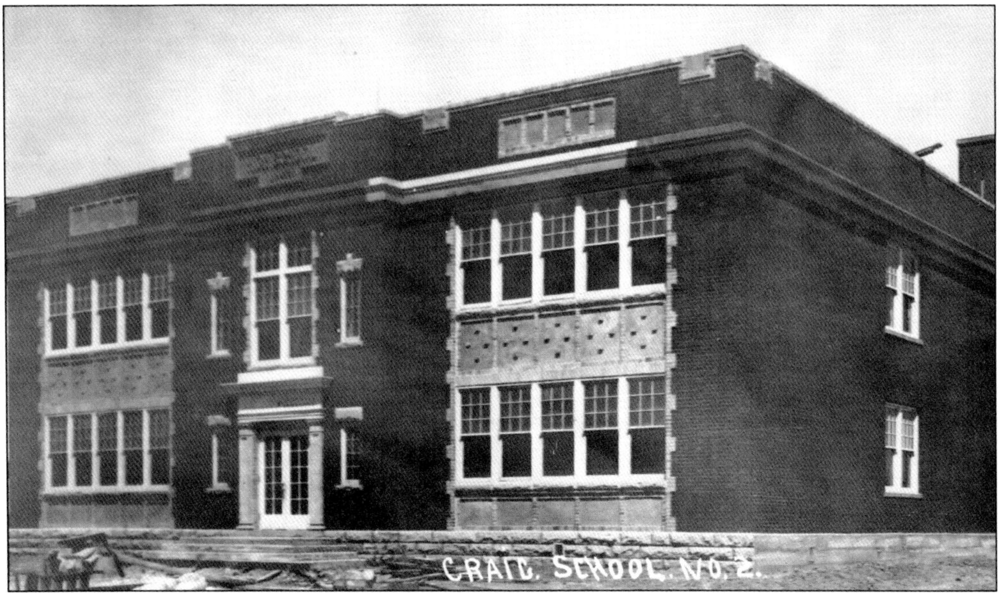

Craig School, No. 2. By 1920, Craig had grown to a population of 1,297, and enrollment at the still new Breeze School was at maximum capacity. Construction began on what would become the new elementary school on Yampa Avenue to accommodate the increased student enrollment. When the building was threatened with demolition in the 1980s, citizens protested and the building was beautifully renovated instead. It now serves as Moffat County's school administration building and a preschool. (Photograph by George Welch.)

Four

EXPANDING HORIZONS

At the end of the Great War, the West had become, for many, a smaller place. Innovations in travel and communication had shrunk the vast distances between frontier communities and the outside world. Young Moffat County war veterans, many of whom had never been out of the valley, returned with expanded horizons and probably, a renewed appreciation for the extraordinary area they called home.

The town was still adjusting to the growth spurt caused by the previous decade's influx of homesteaders. With a burst of patriotism and optimism, a national committee had formed that was the catalyst for the creation of the Victory Highway, which would link the country from coast to coast. Coming right through Craig, the highway would open the doors to continental automobile travel.

Soon, an oil discovery would result in the construction of a refinery in Moffat County. This would be the first of several oil booms to affect the region. Additionally, towards the end of the decade, rail travel would become easier and more reliable due to an incredible engineering feat in the form of a tunnel through the heart of the Rocky Mountains.

The Roaring Twenties were indeed busy ones for Craig, but it roared with the establishment of new businesses and enterprises, new civic and religious buildings, and new growth. At the close of the decade, the town was well established and had adequate resources to weather the biggest challenge it had yet faced: the Great Depression. It would weather this national calamity fairly well, due in part to the solid growth it had experienced in the first part of the 1920s.

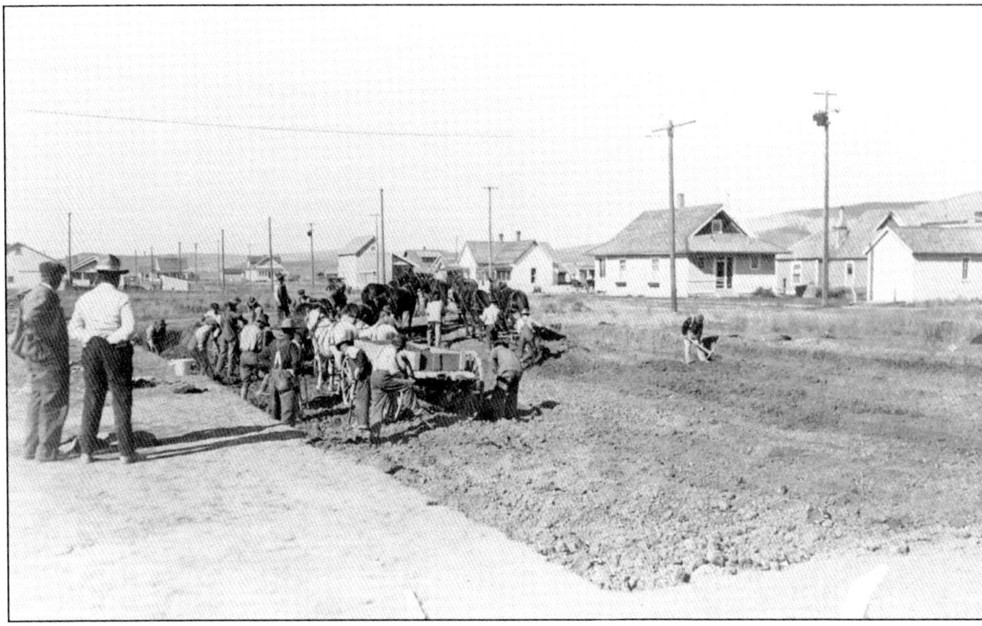

BREAKING GROUND. At the conclusion of the Great War, Craig was poised to move forward again. The townsmen had joined the National Guard, and in 1921, they began preparing for the construction of a new armory. Despite technological advances in many areas, preparation of the lot at Sixth Street and Yampa Avenue required a lot of men and horsepower.

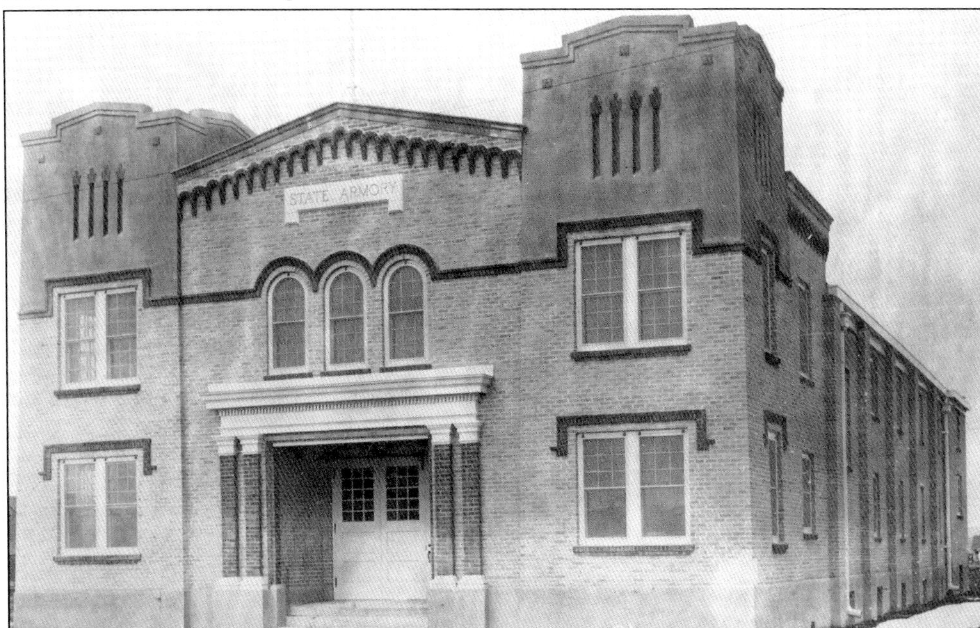

NEW STATE ARMORY. The new armory, completed in 1922, was designed not only to house the National Guard but also to serve as a community building. It boasted a full stage and had a swimming pool in the basement. Today, the beautifully restored structure houses the Museum of Northwest Colorado and also serves the community for meetings and events. (Photograph by George Welch.)

FUNERAL AT THE ARMORY. Joseph Flack, a Spanish-American War veteran and well-liked Craig resident, died in April 1925, and his funeral was held in the armory. Flack, who had owned a local restaurant, also served as justice of the peace, police magistrate, and was the custodian for the armory at the time of his death. (Photograph by George Welch.)

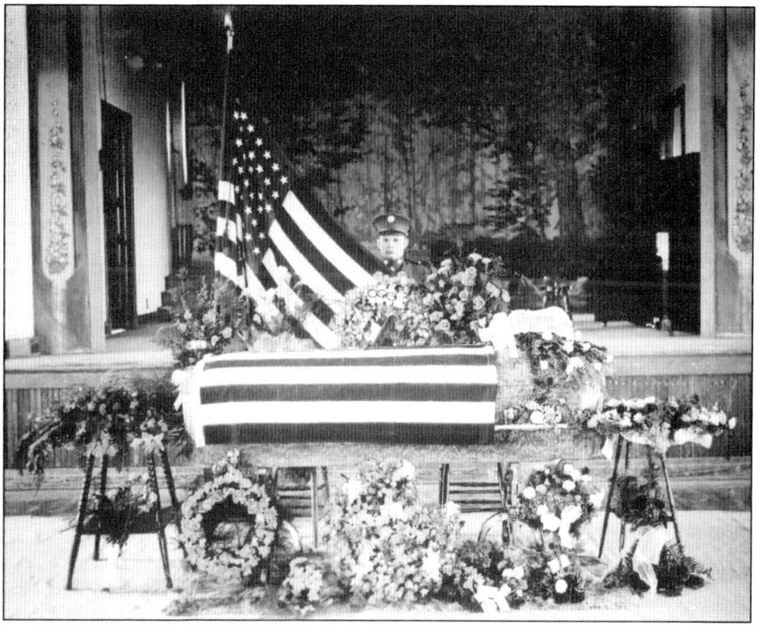

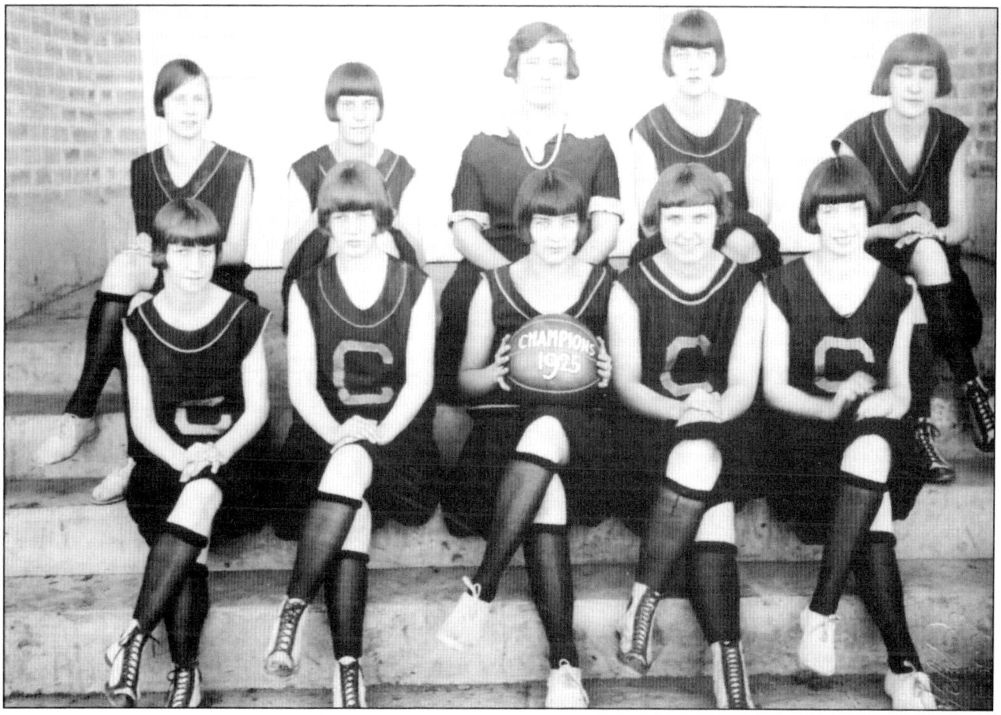

VALLEY CHAMPIONS, 1925. The armory served the community as a sports complex for many decades. Here, the Craig High School Girls Basketball Team poses on the steps of the armory. Players are, from left to right, (first row) Kathryn Finley, Irene Deakins, Eva Miller, Sylvia House, and Erma DeLong; (second row) Billie Irwin, Eva Gulick, Coach Mrs. Roberts, Elva Deakins, and Ruth Thorpe. (Courtesy George Welch.)

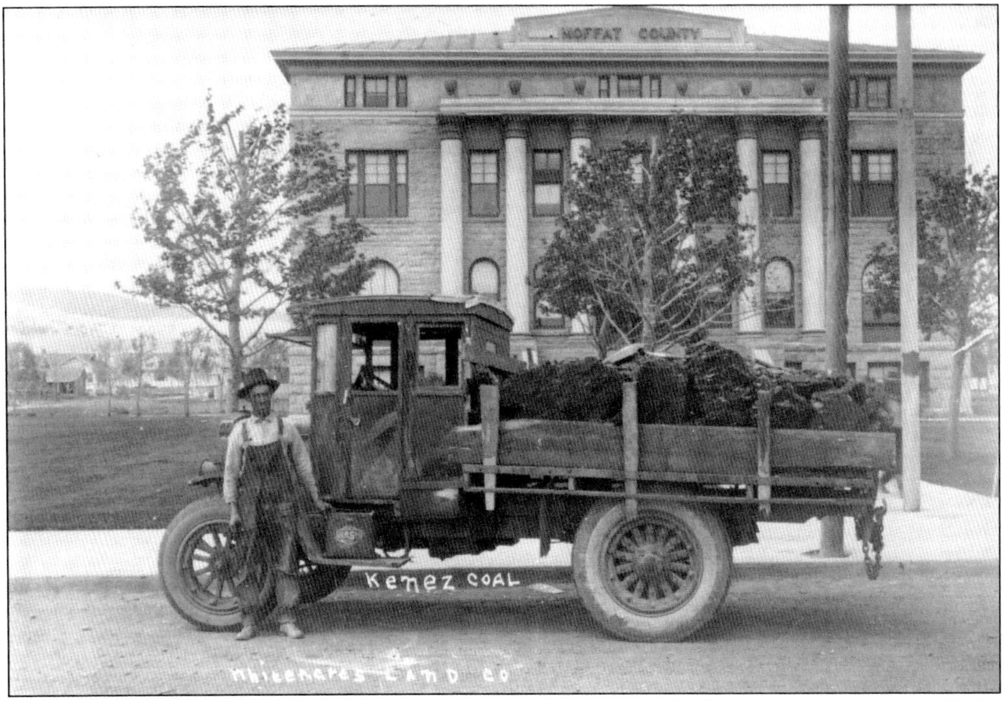

Coal Delivery, 1920s. Family businesses were a mainstay of the local economy in the 1920s. For many years, Joe Knez supplied heating coal to businesses and residences from his mine south of town. At first, he delivered coal with horse-drawn vehicles, but as transportation changed, Joe kept up with the times. Here, Knez poses in front of the courthouse with his Nash truck and a load of coal. (Photograph by George Welch.)

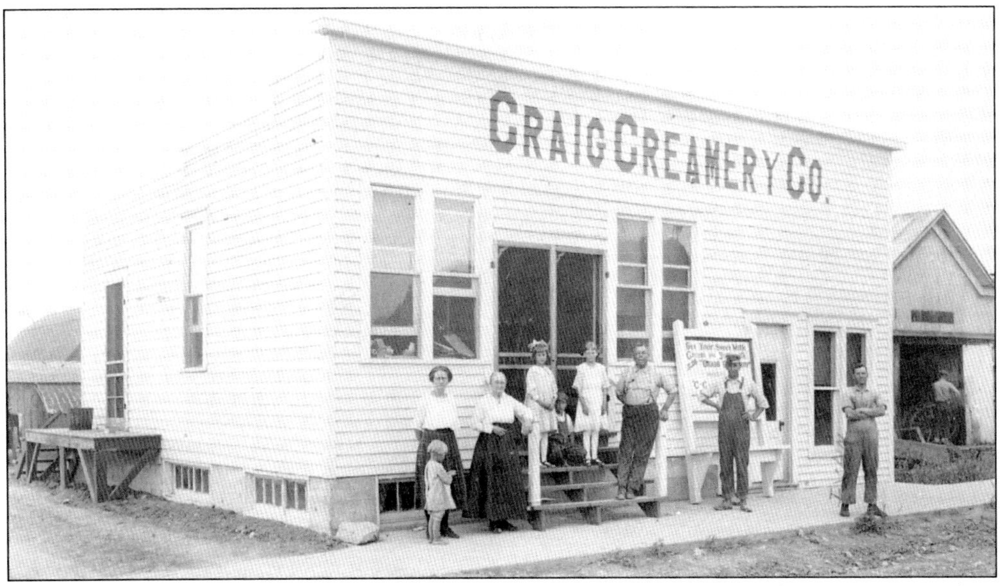

Craig Creamery, 1922. It took the entire Steele family, and some hired help, to operate their creamery on Breeze Street. The Steele children pose with their mother, Pearl (left), grandmother, and father, William, at right on the steps. Claude and Ivan Stevenson stand to the right. This creamery was in operation at this site until the late 1970s. (Photograph by George Welch.)

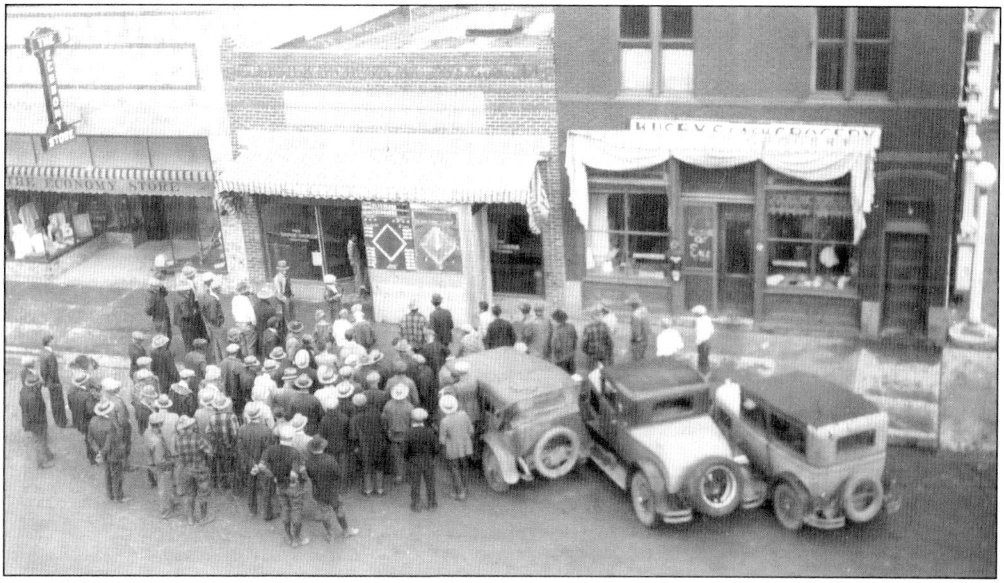

HUB OF CRAIG. The downtown business district often served as a gathering place for all types of events. Men gathered in front of the *Craig Empire Courier* office to watch the play-by-play moves of the 1930 World Series baseball game. The plays, coming through on a radio broadcast, were recorded on the simulated ball field in the window for the viewers. (Photograph by J.N. Bridges.)

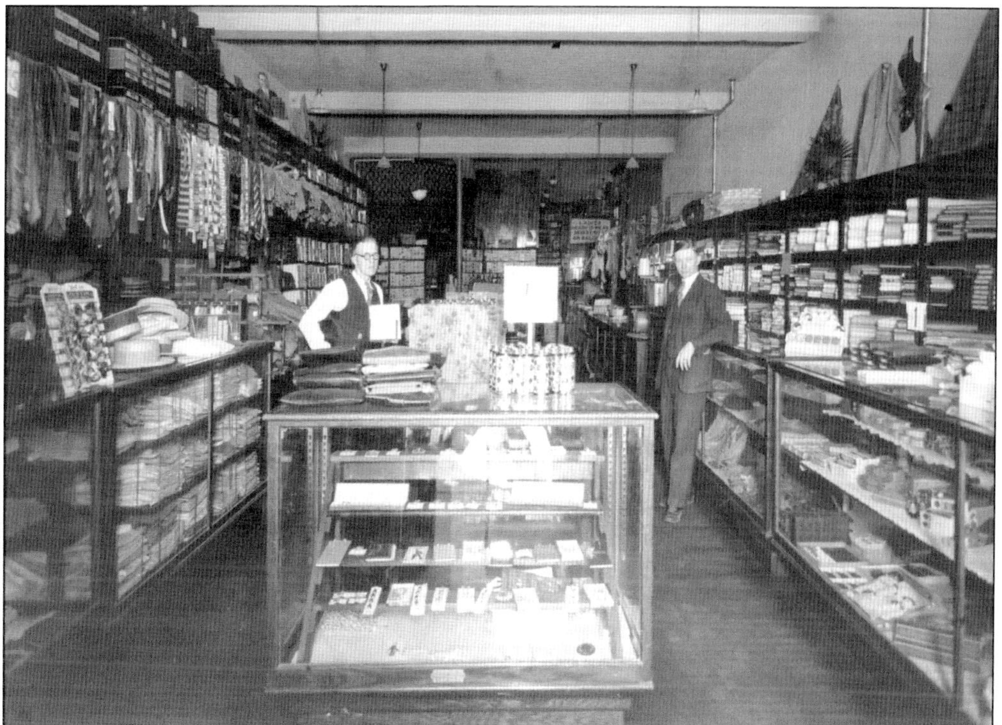

GOLDEN RULE STORE, C. 1930. This family-run store, previously the Economy Store, was located at 506 Yampa. Frank Lindsay, the owner, is on the right. (Photograph by George Welch.)

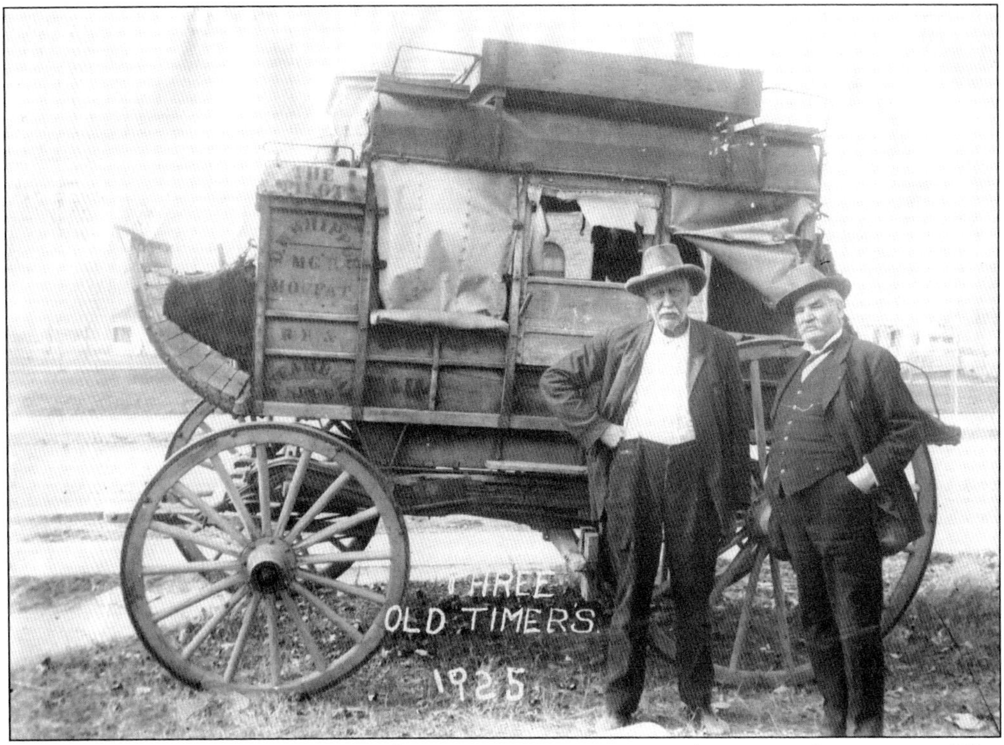

Three Old Timers, 1925. By 1925, horse conveyances were rarely seen on Craig streets. This stage had originally carried passengers on the wagon road between Steamboat Springs and Craig but had been retired for a number of years. Wash Held, also retired, had served as Craig's justice of the peace. C.L. Cleaver managed Craig's first power plant—a municipal electric light plant powered by coal-fired steam. (Photograph by George Welch.)

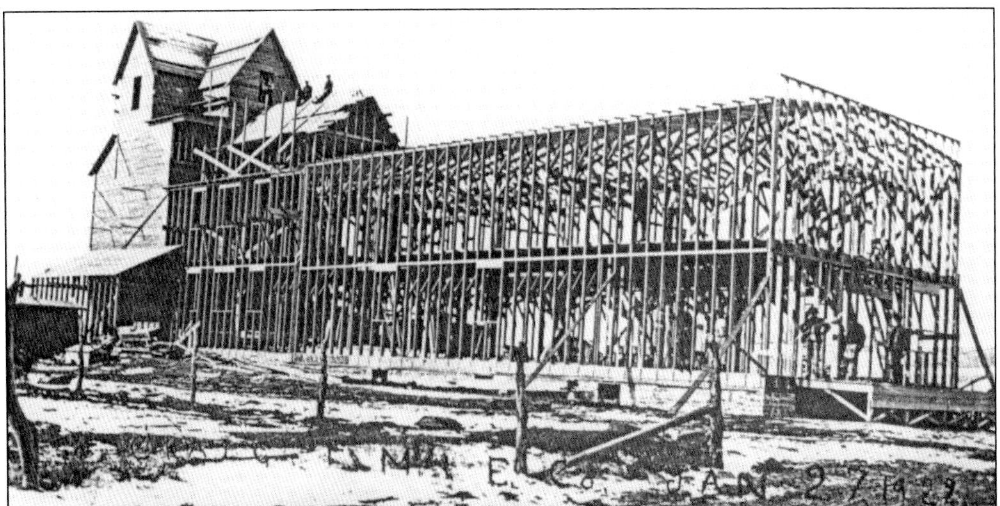

Constructing the Mill, 1922. This mill, located on the west side of Ranney Street, was built to replace the mill that had burned the previous November at the same site. This new mill and elevator served the local farmers until it burned in the 1930s and was again rebuilt.

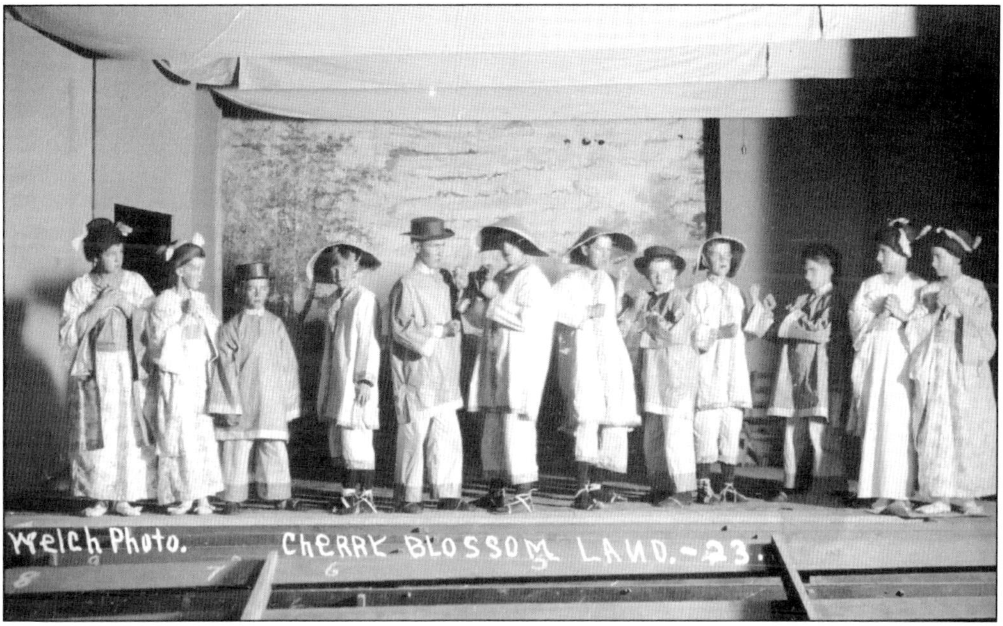

Cherry Blossom Cowboys. The tradition of school plays has a long history worldwide, and Craig schools are no exception. In this 1923 production of *Cherry Blossom Land*, some of the erstwhile Chinamen bear the distinct look of up-and-coming belligerent cowboys. (Photograph by George Welch.)

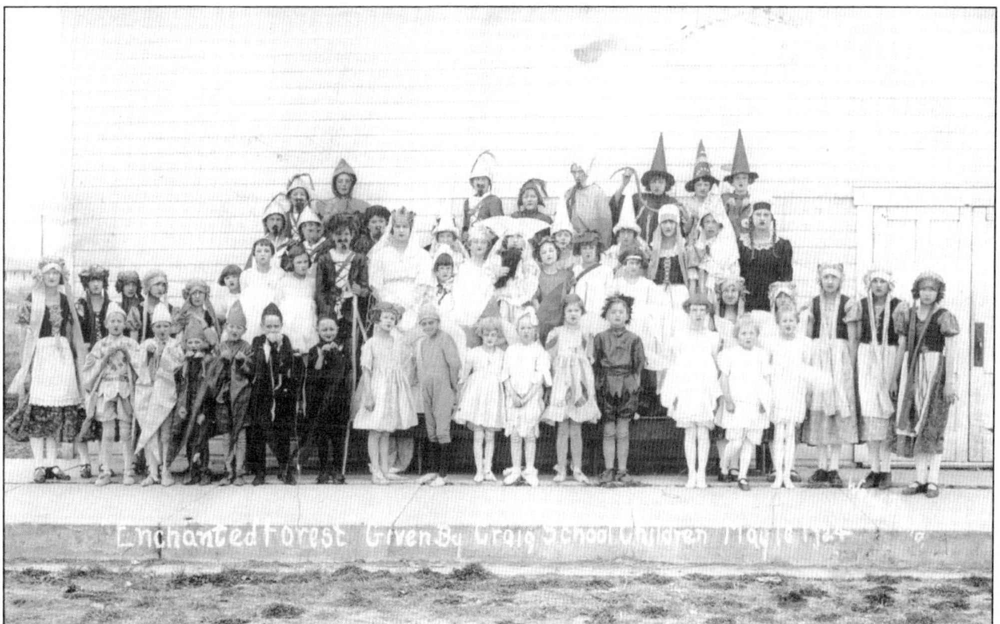

Enchanted Forest. This rather ambitious production appeared to have included all of the elementary children in the Craig school and was performed for the parents and the general town populace in May 1924. The cast is posed outside of Craig's first movie house on Russell Street. That building now is owned by Shepherd & Sons Plumbing and Heating. (Photograph by George Welch.)

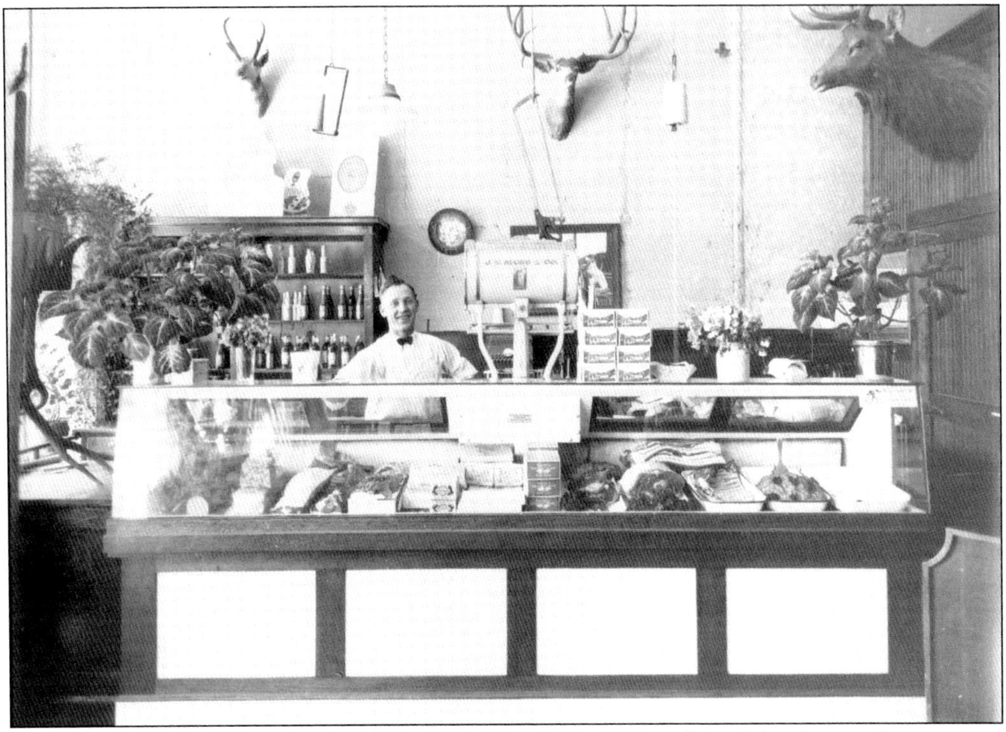

Center Grocery. Just one of several grocery stores in Craig in the 1920s, Center Grocery and Market was located just west of the bank at 13 West Victory Way. Frank Nutter stands ready to serve with a smile behind the meat cases. Note the mounts on the wall solemnly overlooking the entire counter. (Photograph by George Welch.)

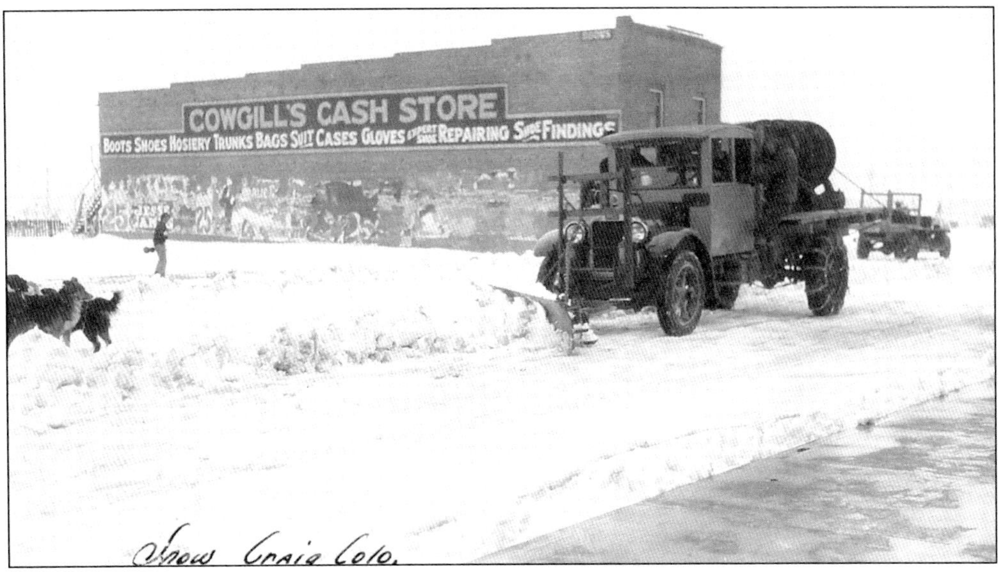

Street Cleaners c. 1925. Under the careful surveillance of a canine crew (left), a plow truck works to clean Yampa Avenue in front of the Cowgill Cash Store after a heavy snowfall. When automobiles first came into use in Craig, residents usually put them in storage for the winter and took to walking. As more cars came into use, plowing snow from the streets was expected.

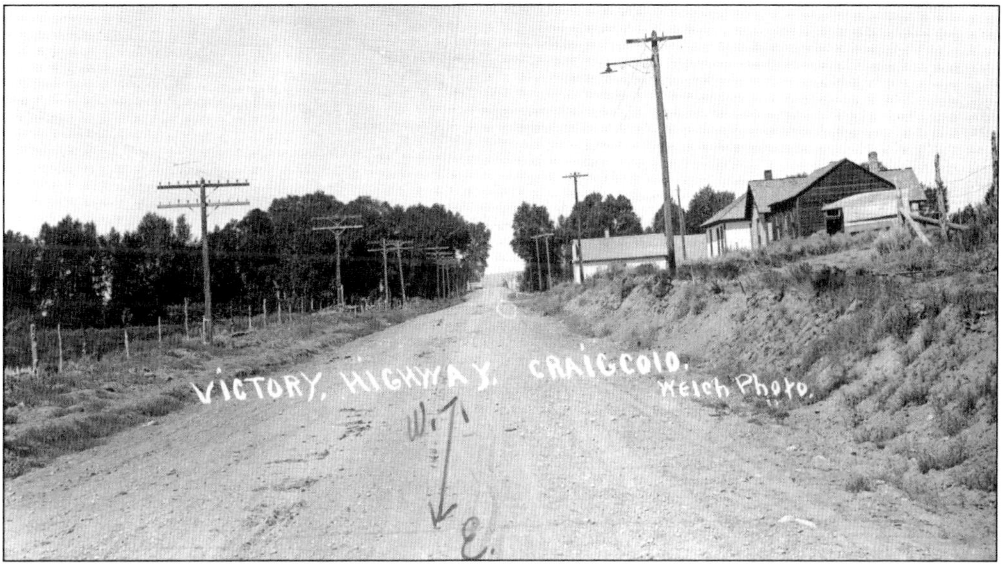

VICTORY HIGHWAY. The Victory Highway Association formed in 1921 to develop a highway that would traverse the country from east to west. Called the Victory Highway in honor of those who had died in the Great War, it was the first road in the United States to receive federal funds for construction. Its route traveled along Main Street in Craig, and the town renamed that street Victory Way to coincide with the highway name. This photograph looks west from the area of today's fairgrounds. (Photograph by George Welch.)

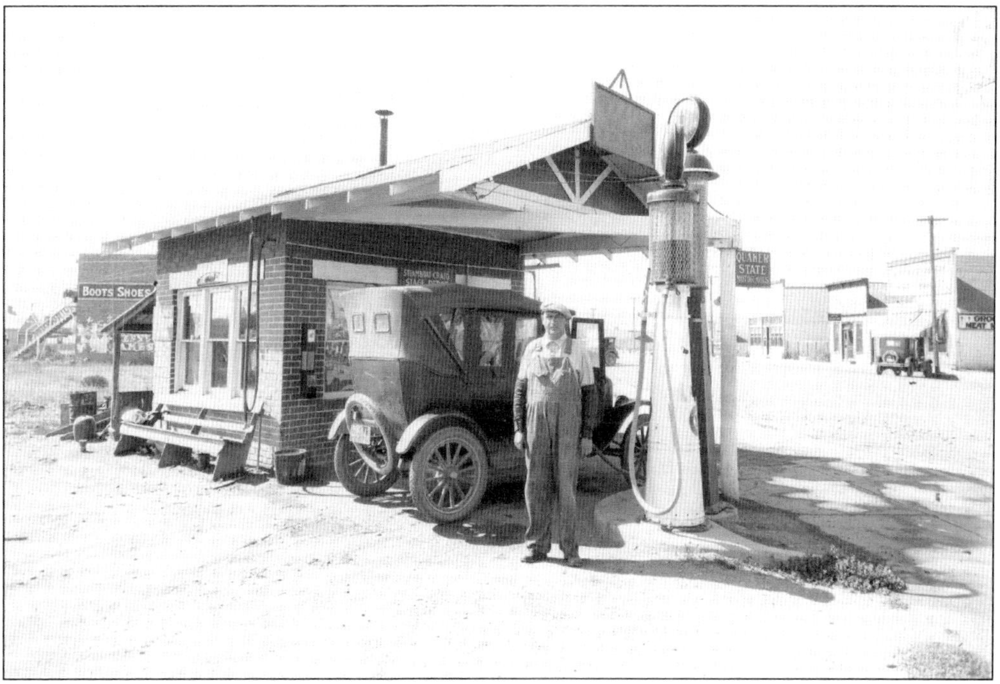

CRAIG GAS STATION, C. 1927. With more people traveling by automobile, a need for venues with gasoline, repair bays, and other automotive necessities developed. Earl Hamilton posed in front of his filling station, which also served as a depot for the Craig-Steamboat stage. It was located on the newly designated Victory Highway at the southeast corner of Victory Way and Yampa Avenue.

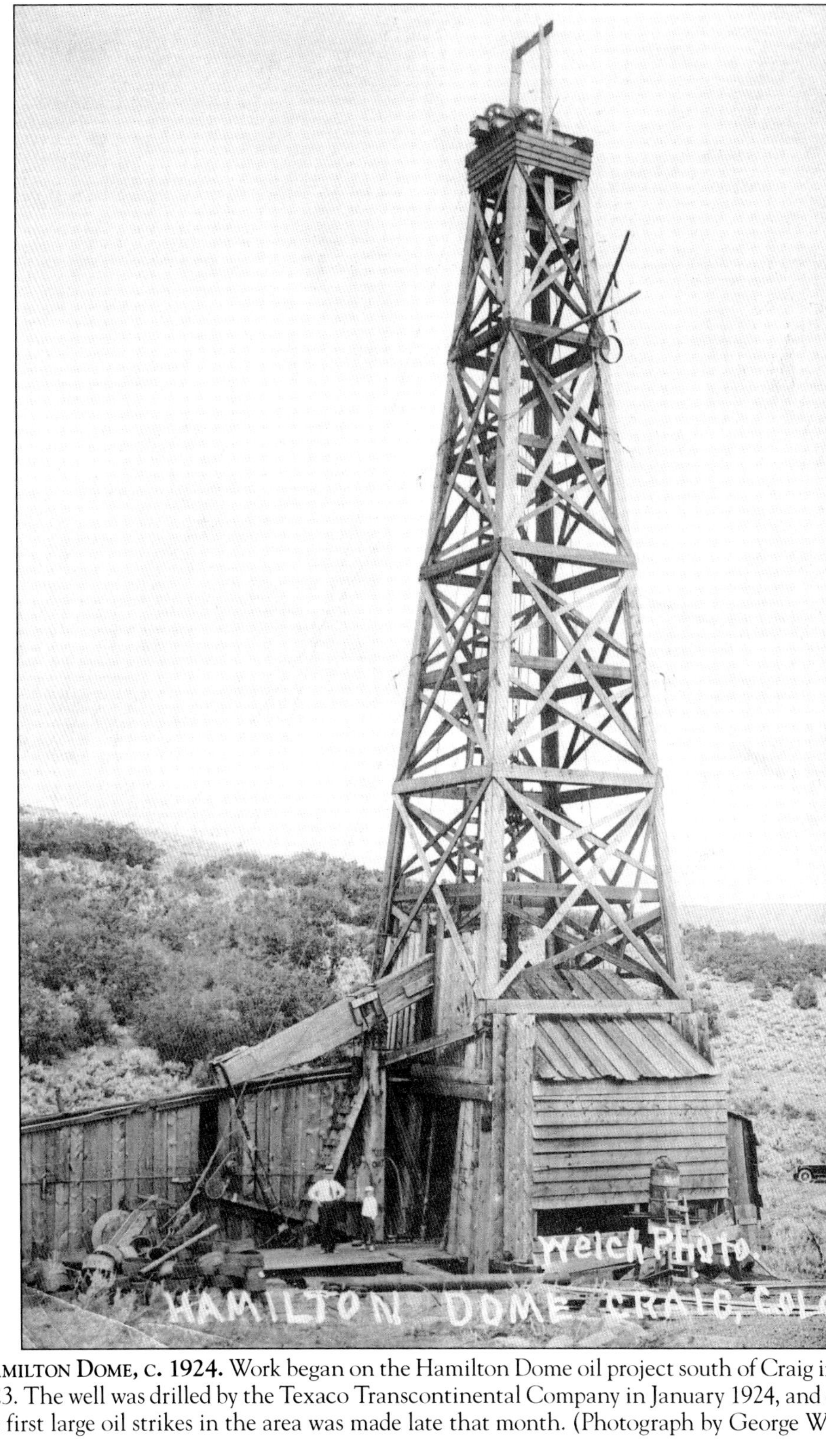

Hamilton Dome, c. 1924. Work began on the Hamilton Dome oil project south of Craig in May 1923. The well was drilled by the Texaco Transcontinental Company in January 1924, and one of the first large oil strikes in the area was made late that month. (Photograph by George Welch.)

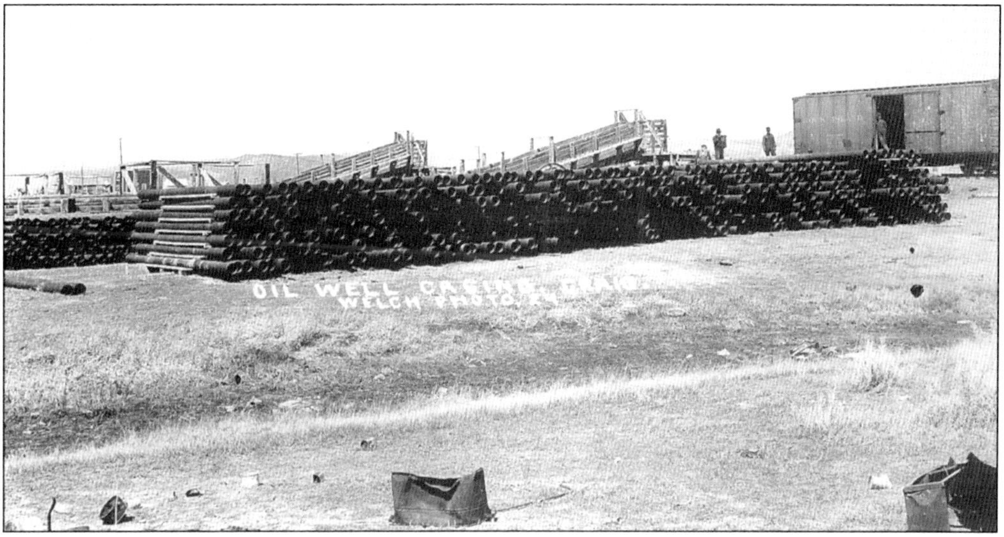

SIGNS OF AN OIL BOOM. Brought in by rail to Craig, oil well casing wait to be delivered to the oil fields south of Craig in 1924. This is a scene that has been repeated several times in Craig's history. (Photograph by George Welch.)

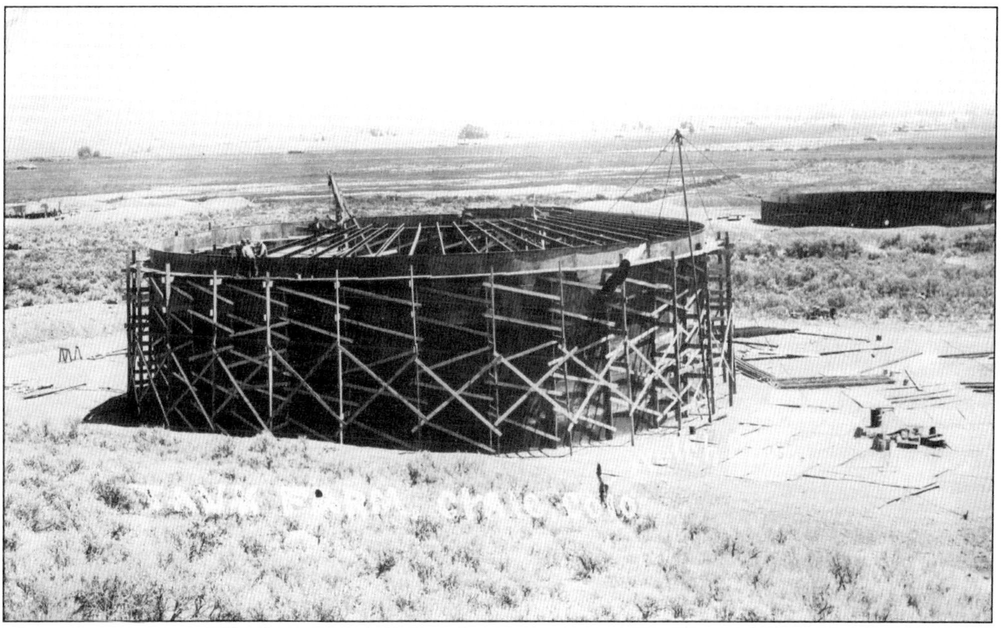

TANK FARM, 1924. The Prairie Oil & Gas Company constructed a tank farm just southwest of Craig with tanks that could store between 35,000 and 55,000 barrels of oil. The oil was piped in 18 miles from the oil fields south of town. (Photograph by George Welch.)

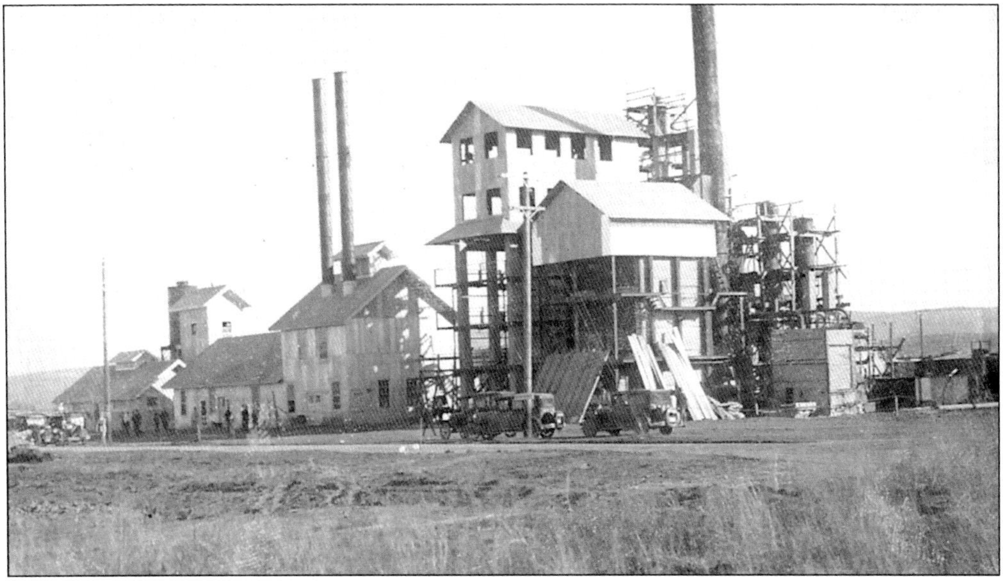

Texaco Oil Refinery in Craig. In 1926 the Texas Company built the refinery west of Craig near present-day Third Street and Highway 13 South. It was a major source of employment for the town during the depression of the 1930s. The refinery closed in 1947 when operations were moved to Casper, Wyoming.

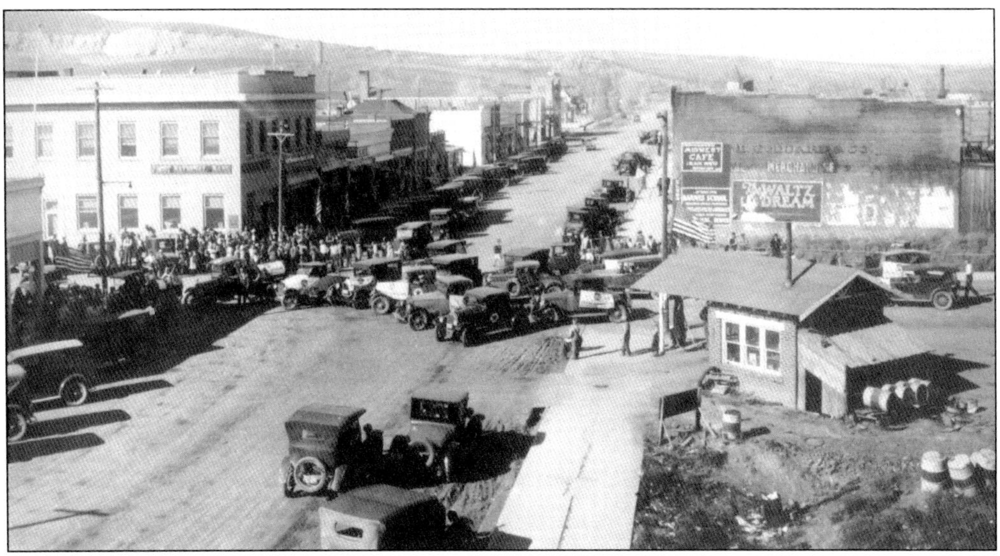

Downtown Celebration. A community celebration was held in honor of the opening of the Texas Refinery in 1926. The view shows the intersection of Yampa Avenue and Victory Way in downtown Craig. The refinery was an important employer for the community for more than 20 years. (Photograph by George Welch.)

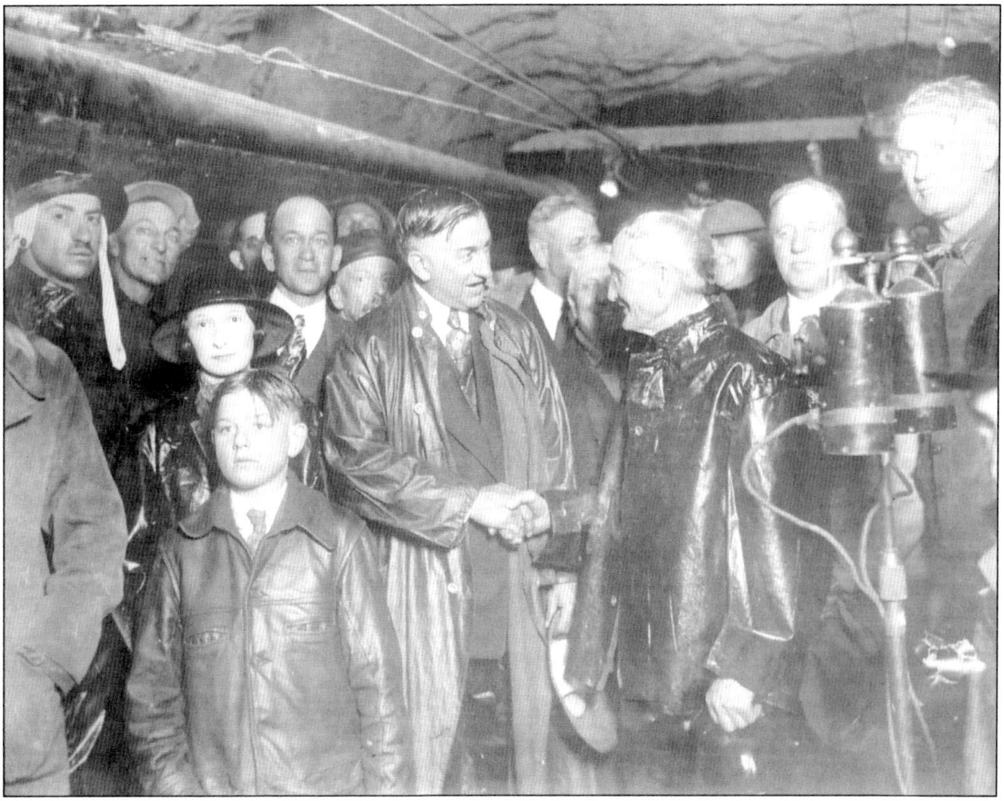

MOFFAT TUNNEL COMPLETION, 1927. Surrounded by construction workers and other well-wishers, Colorado governor Adams congratulates Salt Lake City mayor C.C. Nelsen upon completion of the Moffat Tunnel. This was a momentous occasion marking a new era in the history of the Moffat Road that linked Denver with Craig.

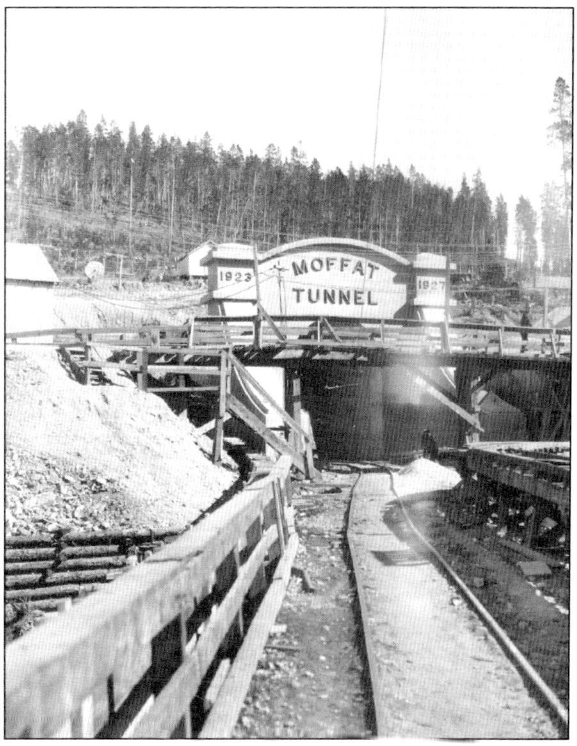

MOFFAT TUNNEL WEST PORTAL, 1927. The newly completed tunnel portal is pictured on the west side. The tunnel carved off several tortuous miles of climbing rails between Craig and Denver, eliminating delays caused by winter snow blockades. It enabled ranchers to more safely transport livestock to Denver, allowed the oil industry to efficiently move its product over the mountains, and improved passenger and mail service.

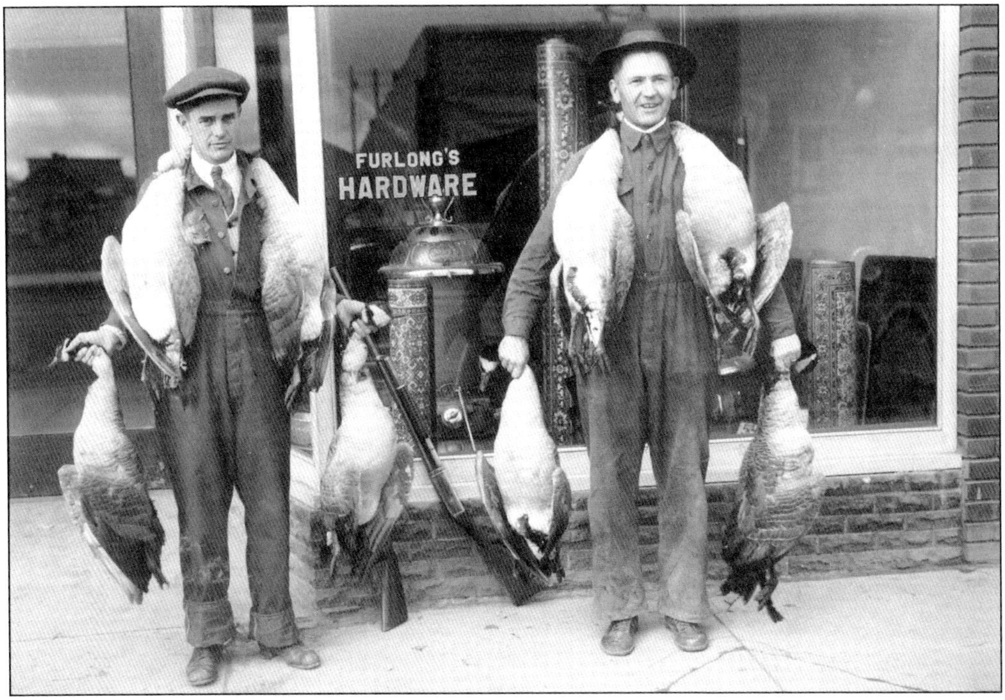

Leisure Time, c. 1925. Alba R. Glassburn (left), manager of Furlong's Hardware, and Judge Charles E. Herrick (right) are pictured after a successful goose hunt. From the beginning, hunting was an important component of the community's identity and still remains a popular activity today. (Photograph by George Welch.)

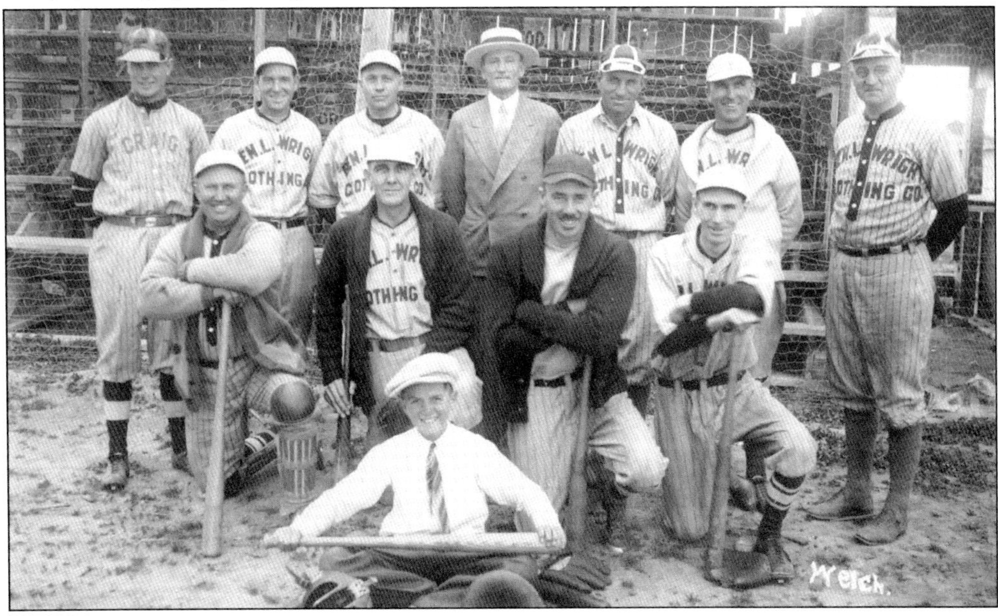

Play Ball, 1927. The Ben L. Wright Clothing Store of Craig sponsored this baseball team. From left to right are (first row) mascot Donald Curtis; (second row) Ed Luttrell, G. Walker, Curtis Chapman, and Loyd DeuPree (third row) Ken Kendall, E. Albritton, C.A. Stoddard, Ben Wright (sponsor), Tex Schiller, Jimmy Jones, and Charles Jandos. (Photograph by George Welch.)

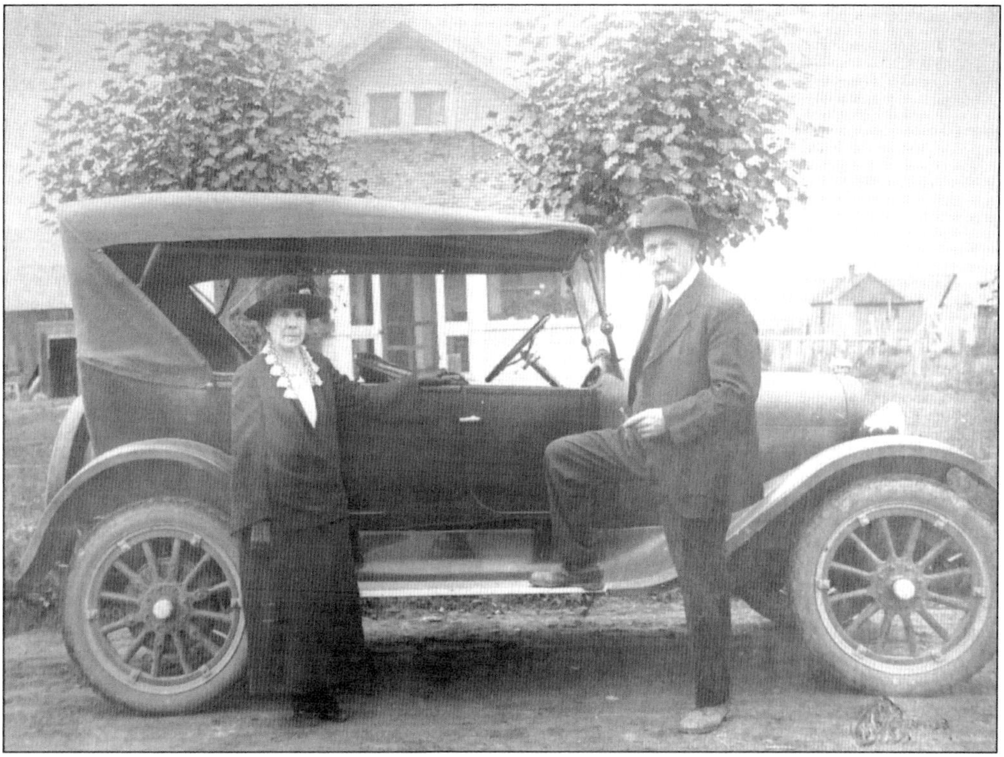

JOHN AND ELLA LEDFORD, C. 1920S. The Ledfords posed alongside their Dodge touring car in front of their new home at 652 Taylor Street. John was the first sheriff of Moffat County, and Ledford Street takes its name from him. Ella was the mother of longtime resident Cullie Walsh. (Photograph by George Welch.)

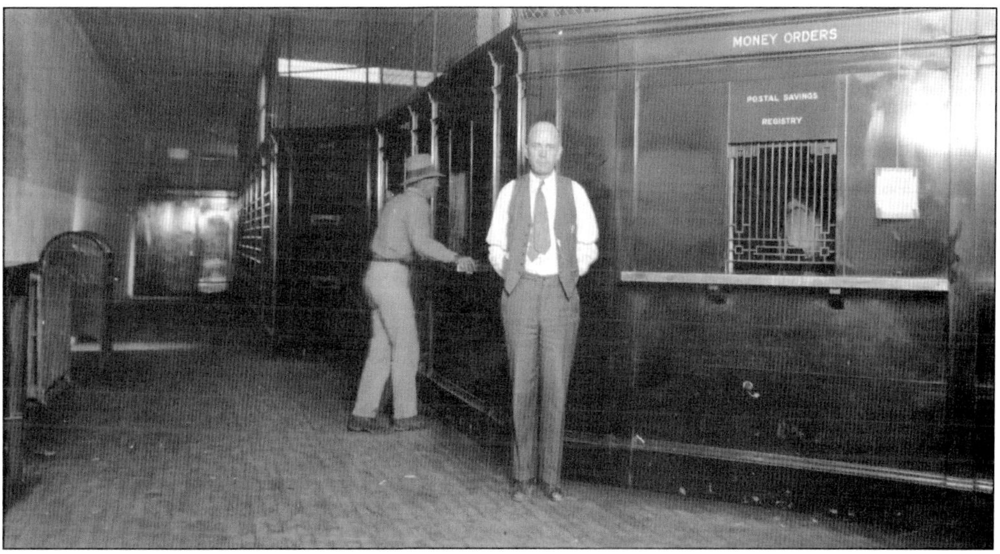

GOVERNMENT BUSINESS, C. 1925. I.P. "Cap" Beckett served as Craig's postmaster from 1920 to 1935. Here, he posed inside the post office building located at 465 Yampa Avenue. This building served as the post office from 1925 until 1953, when it moved to a new building at 552 Russell Street.

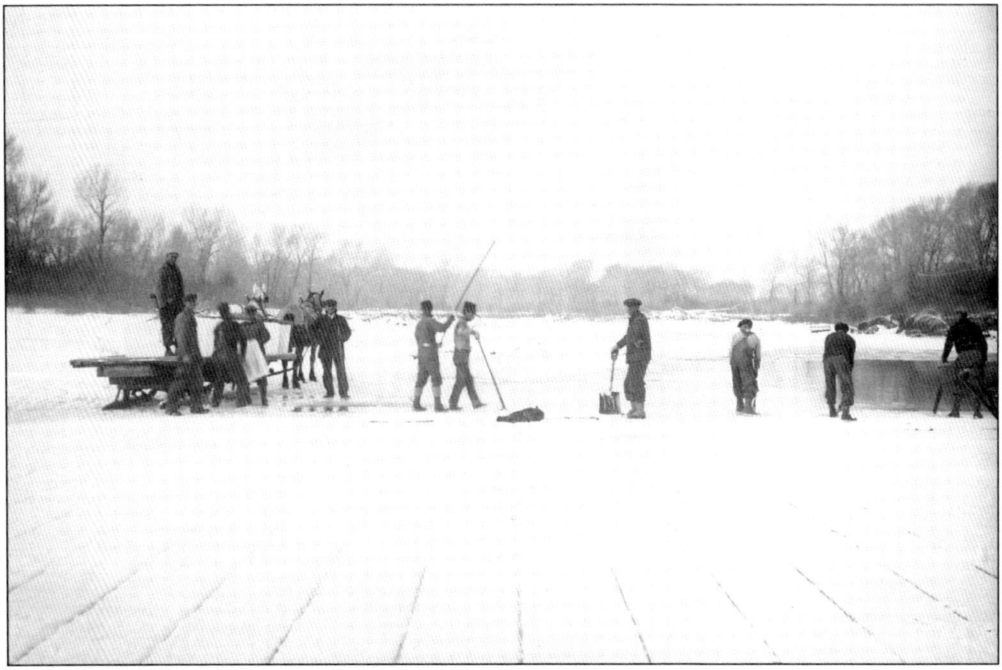

KEEPING THINGS COOL, c. 1930. Ice harvesting was a common practice until electric and gas refrigeration made it obsolete. These c. 1930 photographs show Harry Hansen's crew cutting and loading ice at the Yampa River south of town. The ice was then stored in a warehouse until warmer days, when it was used to cool iceboxes. (Photograph by George Welch.)

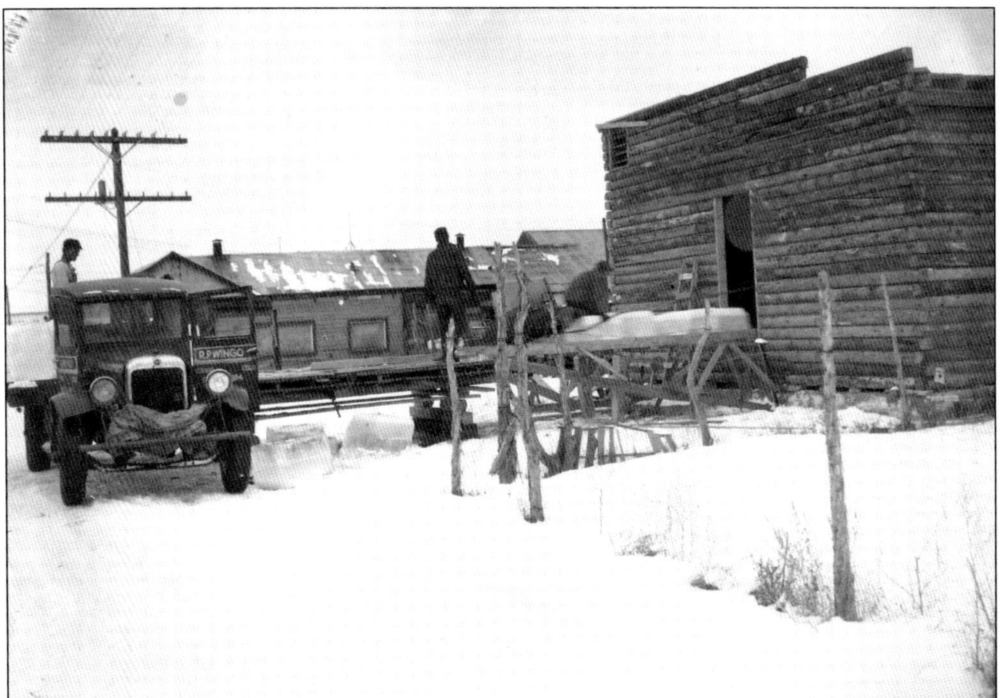

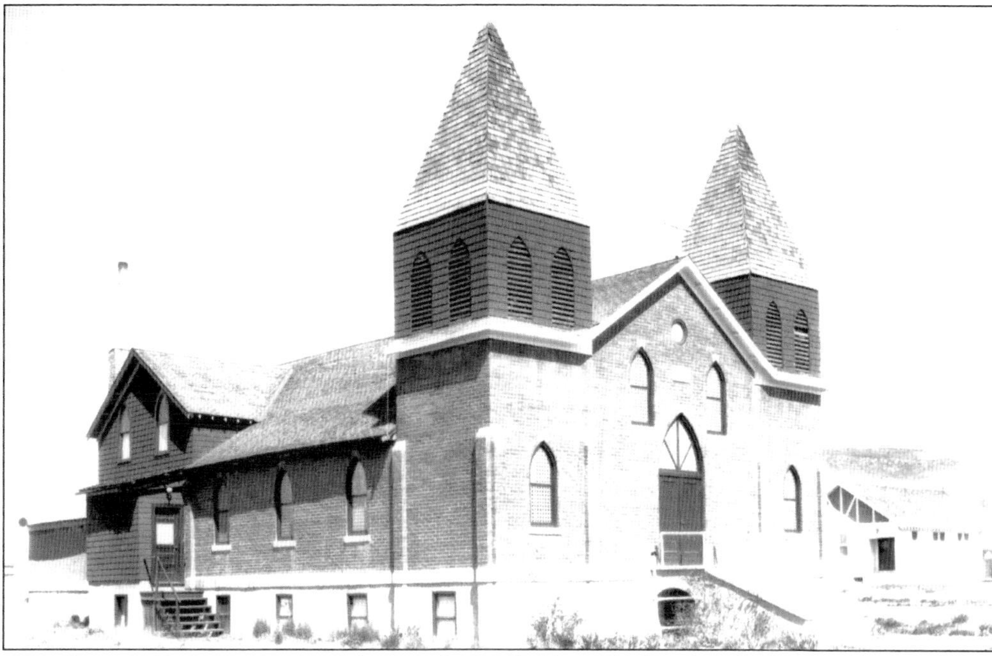

NEW CHURCH BUILDING, 1925. The new St. Michael's Catholic Church is pictured shortly after completion in 1925. The church, located in the 600 block of School Street, served the congregation until 1983, when a new building was erected. (Photograph by George Welch.)

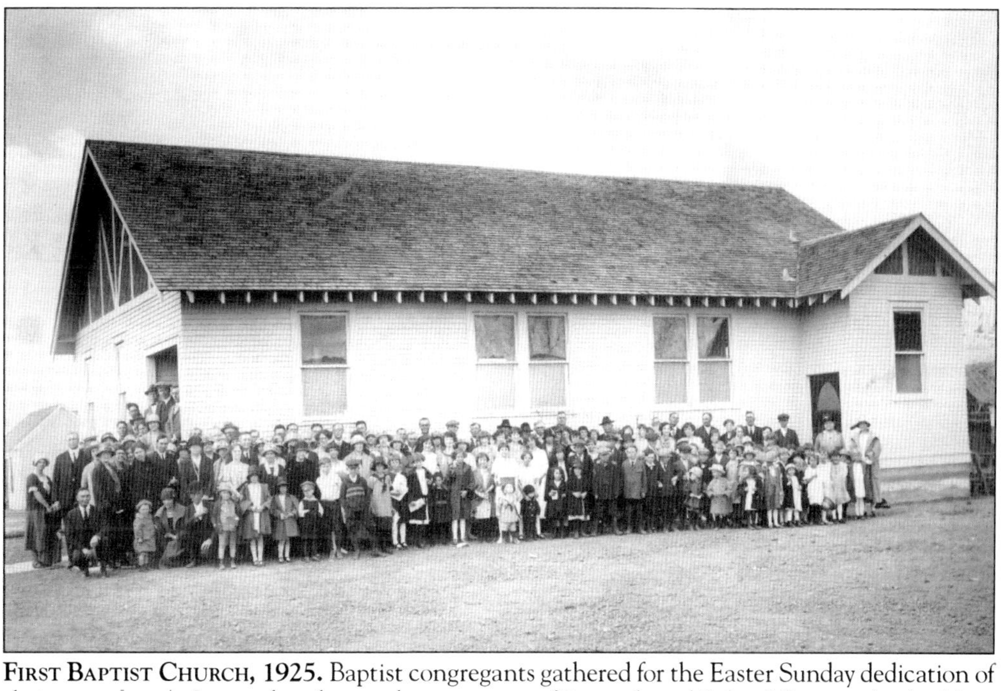

FIRST BAPTIST CHURCH, 1925. Baptist congregants gathered for the Easter Sunday dedication of their new church. Located at the northwest corner of Seventh and School Streets, this building was completed about the same time as the Catholic church on the same street. (Photograph by George Welch.)

Discover Thousands of Local History Books Featuring Millions of Vintage Images

Arcadia Publishing, the leading local history publisher in the United States, is committed to making history accessible and meaningful through publishing books that celebrate and preserve the heritage of America's people and places.

Find more books like this at
www.arcadiapublishing.com

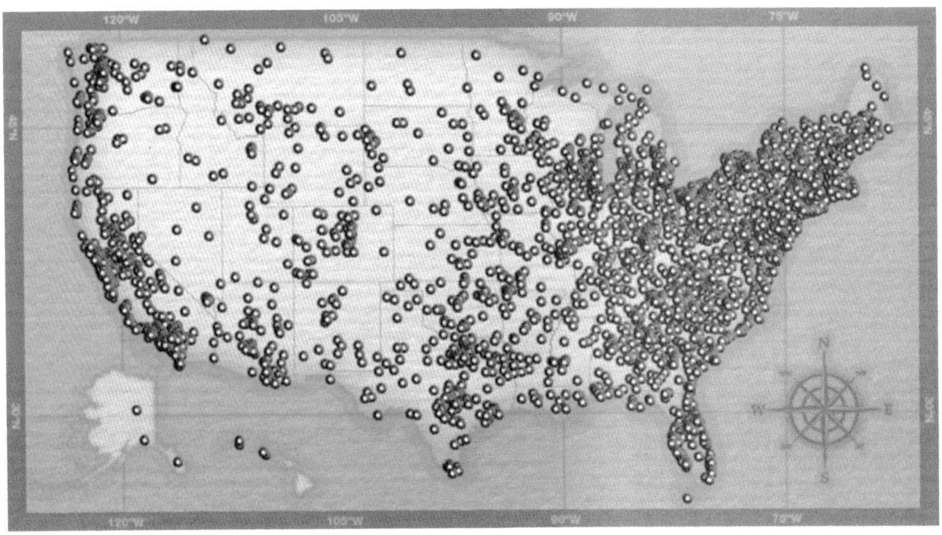

Search for your hometown history, your old stomping grounds, and even your favorite sports team.

Consistent with our mission to preserve history on a local level, this book was printed in South Carolina on American-made paper and manufactured entirely in the United States. Products carrying the accredited Forest Stewardship Council (FSC) label are printed on 100 percent FSC-certified paper.

MADE IN THE USA